Praise for *Regenesis*

"In *Regenesis* . . . George Church and Ed Regis imagine a world where micro-organisms are capable of producing clean petroleum or detecting arsenic in drinking water, where people sport genetic modifications that render their bodies impervious to the flu, or where a synthetic organism can be programmed to invade and destroy cancer cells."

—*Wall Street Journal*

"The life sciences emerge as the new high-tech in this paean to synthetic biology. . . . Each step in the genome's evolution serves as a springboard for expositions of how synthetic biology will revolutionize renewable energy, multivirus resistance, and more." —*Nature*

"Bold and provocative . . . Church and Regis offer a behind-the-scenes look at synthetic biology, a rapidly emerging field that is reprogramming the genetic code to create organisms and functions not found in nature. *Regenesis* tells of recent advances that may soon yield endless supplies of renewable energy, increased longevity, and the return of long-extinct species. . . . Thought-provoking." —*New Scientist*

"[A] phenomenal read." —io9

"If there's one book that can turn this movement into a full-blown revolution, this is it." —O'Reilly Radar

"[*Regenesis*] is a fantastic book and essential reading for anyone interested in the future of science and humanity. It is like a new *Engines of Creation* in the boldness of its goals but with near term objectives that already have companies and labs working towards the goals." —Next Big Future

"[Church and Regis] combine science history and futurism in an exploration of how genomic modification will change the world."

—Zócalo Public Square

"One of the more important new science-related books you can read right now." —Greg Laden, Science Blogs

"Exhilarating and scary facts suffuse this book about bioengineering by leading Harvard genetics professor and entrepreneur Church. . . . When Church describes current work building microbes with minimal genes, the book takes off—and eventually soars. . . . [A] stimulating book."

—*Publishers Weekly*

"Geneticist Church and science writer Regis take a novel evolutionary approach to explaining the science of synthetic biology. . . . [A] highly readable book."

—*Choice*

"[An] authoritative, sometimes awe-inspiring book. . . . A valuable glimpse of science at the edge."

—*Kirkus Reviews*

"A thoughtful introduction to one of the great frontiers of science, one with the promise of literally saving the world. . . . Engaging, readable, and thoroughly fascinating."

—Steven Pinker, Harvard College Professor of Psychology,
Harvard University, and author of *How the Mind Works*
and *The Better Angels of Our Nature*

"*Regenesis* is the most compelling bit of prophecy since the Old Testament first came out in hardback."

—Misha Angrist, Assistant Professor, Duke Institute for
Genome Sciences & Policy and author of *Here is a Human Being*

"A delightfully opinionated, visionary and controversial romp through synthetic biology, which is one of the most important technologies of our time."

—Nathan Myhrvold, Founder and CEO, Intellectual Ventures

"Here you will find the bleeding, screaming, thrilling edges of what is becoming possible with genomic engineering, handsomely framed in the fine-grained fundamentals of molecular biology. It is a combination primer and forecast of what is coming in this 'century of biology' from the perspective of a leading pioneer in the science."

—Stewart Brand, author of *Whole Earth Discipline*

REGENESIS

*How Synthetic Biology Will
Reinvent Nature and Ourselves*

GEORGE CHURCH & ED REGIS

BASIC BOOKS
A Member of the Perseus Books Group
NEW YORK

Books published by Basic Books are available at special discounts for bulk
purchases in the United States by corporations, institutions, and other
organizations. For more information, please contact the Special Markets
Department at the Perseus Books Group, 2300 Chestnut Street, Suite 200,
Philadelphia, PA 19103, or call (800) 810-4145, ext. 5000, or e-mail
special.markets@perseusbooks.com.

A grant from the Alfred P. Sloan Foundation supported the research and
writing of this book.

Designed by Timm Bryson

The Library of Congress has cataloged the hardcover edition as follows:
Church, George M. (George McDonald)
  Regenesis : how synthetic biology will reinvent nature and ourselves / George
M. Church and Ed Regis. — 1st ed.
    p. cm.
  Includes bibliographical references and index.
  ISBN 978-0-465-02175-8 (hardback) — ISBN 978-0-465-03329-4 (e-book) 1.
Synthetic biology. 2. Genomics—Social aspects. 3. Genetics—Social aspects. 4.
Nature. 5. Bioengineering. I. Regis, Edward, 1944- II. Title.
  QH438.7.C486 2012
  572.8'6—dc23

                                                              2012013274
ISBN 978-0-465-07570-6 (paperback)

10 9 8 7 6 5 4 3 2

GEORGE CHURCH
*dedicates this book to his family and to his colleagues
who have been so very supportive of technology
development and regenerative biology.*

ED REGIS
dedicates this book to his wife, Pamela Regis.

CONTENTS

Contents

From Bioplastics to
H. Sapiens 2.0

In December 2009, patrons of the John F. Kennedy Center for the Performing Arts in Washington, DC, experienced a mild jolt of biological future shock when their pre-performance and intermission drinks—their beers, wines, and sodas—were served to them in a new type of clear plastic cup. The cups looked exactly like any other transparent plastic cup produced from petrochemicals, except for a single telling difference: each one bore the legend, "Plastic made 100% from plants."

Plants?

Indeed. The plastic, known as Mirel, was the product of a joint venture between Metabolix, a Cambridge, Massachusetts, bioengineering firm, and Archer Daniels Midland, the giant food processing company that had recently constructed a bioplastics production plant in Clinton, Iowa. The plant had been designed to churn out Mirel at the rate of 110 million pounds per year.

Chemically, Mirel was a substance known as polyhydroxybutyrate (PHB), which was normally made from the hydrocarbons found in petroleum. But starting in the early 1990s, Oliver Peoples, a molecular biologist who was a cofounder of Metabolix, began looking for ways to produce polymers like PHB by fermentation, by the action of genetically altered microbes on a feedstock mixture.

After seventeen years of research and experimentation (and having been laughed out the doors of several chemical companies), Peoples had developed an industrial strain of a proprietary microbe that turned corn sugar into the PHB plastic polymer. In its broadest outlines, the process was not all that different from brewing beer, which was also accomplished by fermentation: microorganisms (yeast cells) acted on malt and hops to produce ethanol. In the case of Mirel, the microbial fermentation system consisted of a large vat that combined the engineered microbes with corn sugar and other biochemical herbs and spices. The microbes metabolized the corn sugar and turned it into bioplastic, which was then separated from the organisms and formed into pellets of Mirel. Ethanol was a chemical, and so was PHB, but in both cases microbes effected the transformation of organic raw material into a wholly different kind of finished product.

The microbial-based PHB had some key environmental advantages over the petrochemical-derived version. For one thing, since it wasn't made from petroleum, it lessened our dependence on fossil fuels. For another, its chief feedstock material, corn, was an agriculturally renewable and sustainable resource, not something we were going to run out of any time soon. For a third, Mirel bioplastic resins were the only nonstarch bioplastics certified by Vinçotte, an independent inspection and certification organization, for biodegradability in natural soil and water environments, such as seawater. If any of the plastic cups used at the Kennedy Center ended up in the Potomac River, they would break down and be gone forever in a matter of months. (Biodegradation is not necessarily the panacea it was once thought to be, since it releases greenhouse gases, while nondegradation, ironically, sequesters carbon.)

Constructing a microbe that would convert corn into plastic, in a process akin to beer brewing, was just one example of the transformations made possible by the emerging discipline of synthetic biology—the science of selectively altering the genes of organisms to make them do things that they wouldn't do in their original, natural, untouched state.

But the feat of turning corn into plastic was merely the tip of the synthetic biology iceberg. By the first decade of the twenty-first century microbe-made commodities were yielding up products that nobody would have guessed were manufactured by bacteria in three-story-high industrial vats. Carpet fibers, for example.

In 2005 Mohawk Industries introduced its new SmartStrand carpet line. It was based on the DuPont fiber Sorona, which was made out of "naturally occurring sugars from readily available and renewable crops." The Sorona fiber had a unique, semicrystalline molecular structure that made it especially suitable for clothing, automobile upholstery, and carpets. The fiber had a pronounced kink in the middle, and the shape acted as a molecular spring, allowing the strands to stretch or deform and then automatically snap back into their original shape. That attribute was perfect for preventing baggy knees or elbows, or for making carpets that were highly resilient, comfortable, and supportive.

Sorona's main ingredient was a chemical known as 1,3-propanediol (PDO), which was classically derived from petrochemicals and other ingredients that included ether, rhodium, cobalt, and nickel. In 1995 DuPont had teamed up with Genencor International, a genetic engineering firm with principal offices in Palo Alto, to research the possibility of producing PDO biologically. Scientists from the two companies took DNA from three different microorganisms and stitched them together in a way that resulted in a new industrial strain of the bacterium *Escherichia coli*. Specifically, they programmed twenty-six genetic changes into the microbe enabling it to convert glucose from corn directly into PDO in a fermenter vat, like beer and Mirel.

In 2003 DuPont trademarked the name Bio-PDO and started producing the substance in quantity. The company claimed that this was the first time a genetically engineered organism had been utilized to transform a naturally occurring renewable resource into an industrial chemical at high volumes. The US Environmental Protection Agency, which regarded Bio-PDO as a triumph of green chemistry, gave DuPont the 2003 Greener Reaction Conditions Award (a part of the Presidential Green Chemistry Challenge). And why not? The biofiber used greener feedstocks and reagents, and its synthesis required fewer and less expensive process steps than were involved in manufacturing other fibers. The production of Sorona consumed 30 percent less energy than was used to produce an equal amount of nylon, for example, and reduced greenhouse gas emissions by 63 percent. For its part, Mohawk touted its Sorona carpeting as environmentally friendly: "Every seven yards of SmartStrand with DuPont Sorona saves enough energy and resources to equal one gallon of gasoline—that's 10 million gallons of gasoline a year!" Here it was, finally: the politically correct carpet.

What these examples hinted at, however, was something far more important than mere political correctness, namely, that biological organisms could be viewed as a kind of high technology, as nature's own versatile engines of creation. Just as computers were universal machines in the sense that given the appropriate programming they could simulate the activities of any other machine, so biological organisms approached the condition of being universal constructors in the sense that with appropriate changes to their genetic programming, they could be made to produce practically any imaginable artifact. A living organism, after all, was a ready-made, prefabricated production system that, like a computer, was governed by a program, its genome. Synthetic biology and synthetic genomics, the large-scale remaking of a genome, were attempts to capitalize on the facts that biological organisms are programmable manufacturing systems, and that by making small changes in their genetic software a bioengineer can effect big changes in their output. Of course, organisms cannot manufacture just anything, for like all material objects and processes they are limited and circumscribed by the laws of nature. Microbes cannot convert lead into gold, for example. But they can convert sewage into electricity.

This astonishing capacity was first demonstrated in 2003 by a Penn State team headed by researcher Bruce Logan. He knew that in the United States alone, more than 126 billion liters of wastewater was treated every day at an annual cost of $25 billion, much of it spent on energy. Such costs, he thought, "cannot be borne by a global population of six billion people, particularly in developing countries." It was widely known that bacteria could treat wastewater. Separately, microbiologists had known for years that bacteria could also generate electricity. So far, nobody had put those two talents together. But what if microbes could be made to do both things simultaneously, treating wastewater while producing electrical energy?

Key to the enterprise would be the microbial fuel cell—a sort of biological battery. In ordinary metabolism, bacteria produce free electrons. A microbial fuel cell (MFC) consists of two electrodes—an anode and a cathode. A current is set up between them by the release of electrons from bacteria in a liquid medium. Electrons pass from the bacteria to the anode, which is connected to the cathode by a wire.

Logan and his colleagues constructed a cylindrical microbial fuel cell, filled it with wastewater from the Penn State water treatment plant, and then inoculated it with a pure culture of the bacterium *Geobacter metallireducens*. Lo and behold, in a matter of hours the microbe had begun purifying the sewage while at the same time producing measurable amounts of electricity. These results "demonstrate for the first time electricity generation accompanied by wastewater treatment," Logan said. "If power generation in these systems can be increased, MFC technology may provide a new method to offset wastewater treatment operating costs, making advanced wastewater treatment more affordable for both developing and industrialized nations."

The general setup wasn't difficult to replicate and within a few years a sophomore at Stuyvesant High School in New York City, Timothy Z. Chang, was designing, building, and operating microbial fuel cells at home and in his high school lab. He had experimented with some forty different strains of bacteria to discover which was best suited to maximum electricity production. "It may be possible to achieve even higher power yields through active manipulation of the microbial population," he wrote in a formal report on the project.

By 2010 several teams of researchers were working on scaling up bacterial electricity production from sewage to make it into a practical, working, real-world option. By this time, synthetic biologists had gotten microbes to perform so many different feats of creation that it was clear that many of nature's basic units of life—microbes—were undergoing an extreme DNA makeover, a major course of redesign from the ground up.

Engineered microbes produced diesel oil, gasoline, and jet fuel. Microbes were made to detect arsenic in drinking water at extremely low concentrations (as low as 5 parts per billion) and report the fact by changing color. There were microbes that could be spread out into a biofilm. By producing a black pigment in response to selective illumination, they could copy superimposed patterns and projected images—in effect, microbial Xerox machines.

A student project reprogrammed *E. coli* bacteria to produce hemoglobin ("bactoblood"), which could be freeze-dried and then reconstituted in the field and used for emergency blood transfusions. In 2006, just for fun, five MIT undergrads successfully reprogrammed *E. coli* (which as a resident of the intestinal tract smelled like human waste) to smell like either bananas or wintergreen.

E. coli was so supple, pliable, and yielding that it seemed to be the perfect biological platform for countless bioengineering applications. One of its greatest virtues was that the *E. coli* bacterium (and cousins, the Vibrio) are the world's fastest machines at doubling, small or large.* It reproduced itself every twenty minutes, so that theoretically, given enough simple food and stirring, a single particle of *E. coli* could multiply itself exponentially into a mass greater than the earth in less than two days.

Still, as malleable as it was, University of Wisconsin geneticist Fred Blattner decided he could materially improve the workhouse K-12 strain of the microbe to make it an even better chassis for synthetic biology engineering projects. The microbe had some 4,000 genes; many had no known function, while others were nonessential, redundant, or toxic. So Blattner stripped 15 percent of its natural genes from the K-12 genome, making it a sort of reduced instruction set organism, a streamlined, purer version of the microbe. Blattner described it as "rationally designed" and said that his genetic reduction "optimizes the *E. coli* strain as a biological factory, providing enhanced genetic stability and improved metabolic efficiency." With forty genome changes, he had pre-engineered the microbe in order to make it easier to engineer.

---

* Bacteria called *Clostridium perfringens* and *Vibrio natriegens* seem to be the world's fastest doublers, reproducing in seven to ten minutes respectively.

In 2002 Blattner founded Scarab Genomics to sell his new and improved organism, now billing it as "Clean Genome *E. coli*" and marketing it under the slogan "Less is better and safer!" Researchers can buy quantities of the microbe, online or by fax, for as little as $89 a shot (plus a $50 shipping fee).

The upshot of all this is that, at least at the microbial level, nature has been redesigned and recoded in significant ways. Genomic engineering will become more common, less expensive, and more ambitious and radical in the future as we become more adept at reprogramming living organisms, as the cost of the lab machinery drops while its efficiency rises, and as we are motivated to maximize the use of green technologies.

Given the profusion and variety of biological organisms, plus the ability to reengineer them for a multiplicity of purposes, the question was not so much what they can be made to do but what they can't be made to do, in principle. After all, tiny life forms, driven solely by their own natural DNA, have, just by themselves, produced large, complex objects: elephants, whales, dinosaurs. A minuscule fertilized whale egg produces an object as big as a house. So maybe one day we can program an organism, or a batch of them, to produce not the whale but the actual house. We already have bioplastics that can be made into PVC plumbing pipes; biofibers for carpeting; lumber, nature's own building material; microbe-made electricity to provide power and lighting; biodiesel to power the construction machinery. Why can't other microbes be made to produce whatever else we need?

In 2009 Sidney Perkowitz, a physicist at Emory University in Atlanta with a special interest in materials science, was asked to speculate about the future of building materials. "Think about the science-fictionish possibility of bioengineering plants to produce plastic exactly in a desired shape, from a drinking cup to a house," he said. "Current biotechnology is far short of this possibility, but science fiction has a way of pointing to the future. If bioplastics are the materials breakthrough of the 21st century, houses grown from seeds may be the breakthrough of the 22nd."

Similar proposals have been made by others, and they may be much closer than the twenty-second century; for example, using modified gourds and trees to grow a primitive, arboreal house (inhabitat.com/grow-your-own-treehouse). The technology of determining the shape and chemical properties of plants by making them sensitive to simple cues of light and scaffolding is improving rapidly.

❦   ❦   ❦

This focus on microbes and plants—especially on the overworked *E. coli* bacterium—may give rise to the impression that synthetic biology and genomic engineering have little to offer the charismatic megafauna—the higher organisms such as people. Nothing could be further from the truth. In fact these technologies have the power to improve human and animal health, extend our life span, increase our intelligence, and enhance our memory, among other things.

The idea of improving the human species has always had an enormously bad press, stemming largely from the errors and excesses associated with the eugenics movements of the past. Historically, eugenics has covered everything from selective breeding for the purpose of upgrading the human gene pool to massive human rights violations against classes of people regarded as undesirable, degenerate, or unfit because of traits such as religion, sexual preference, handicap, and so on, culminating, in the extreme case, in the Nazi extermination program.

Some proposals for enhancing the human body have had a harebrained ring to them, as for example the idea of equipping people with gills so that they could live in the sea alongside sharks. Burdened with past evils and silliness, any new proposal for changing human beings through genomic engineering faces an uphill battle. But consider this modest proposal: What if it were possible to make human beings immune to all viruses, known or unknown, natural or artificial? No more viral epidemics, influenza pandemics, or AIDS infections.

Viruses do their damage by entering the cells of the host organism and then using the cellular machinery to replicate themselves, often killing the host cells in the process. This leads to the release of new viruses that proceed to infect other cells, which in turn produce yet more virus particles, and so on. Viruses can take control of a cell's genetic machinery because both the virus and the cell share the same genetic code. However, changing the genetic code of the host cell, as well as that of the cellular machinery that reads and expresses the viral genome, could thwart the virus's ability to infect cells (see Chapter 5).

All this may sound wildly ambitious, but there is little doubt that the technology of genome engineering is in principle up to the task. An additional benefit of engineering a sweeping multivirus resistance into the body is that it would alleviate a common fear concerning synthetic biology—the accidental creation of an artificial supervirus to which humans would have no natural immunity.

Genomic technologies can actually allow us to raise the dead. Back in 1996, when the sheep Dolly was the first mammal cloned into existence, she was not cloned from the cells of a live animal. Instead, she was produced from the frozen udder cell of a six-year-old ewe that had died some three years prior to Dolly's birth. Dolly was a product of nuclear transfer cloning, a process in which a cell nucleus of the animal to be cloned is physically transferred into an egg cell whose nucleus had previously been removed. The new egg cell is then implanted into the uterus of an animal of the same species, where it gestates and develops into the fully formed, live clone.

Although Dolly's genetic parent had not been taken from the grave and magically resurrected, Dolly was nevertheless probably a nearly exact genetic duplicate of the deceased ewe from which she had been cloned, and so in that sense Dolly had indeed been "raised from the dead." (Dolly was certainly different in the details of how the genome played out developmentally [a.k.a. epigenetically] but not so different as to discourage subsequent success in a variety of agricultural and research species.)

But even better things were in the offing. A few years after Dolly, a group of Spanish and French scientists brought to life a member of an extinct animal species—the Pyrenean ibex, or bucardo, a subspecies of wild mountain goat whose few remaining members had been confined to a national park in northern Spain. The species had become extinct in January 2000, when the very last living member, a thirteen-year-old female named Celia, was crushed to death by a falling tree. Consequently the International Union for the Conservation of Nature (IUCN) formally changed the conservation status of the species from EW, which meant "extinct in the wild," to EX, which meant "extinct," period.

Extinction, supposedly, was forever.

But in the spring of 1999, Dr. Jose Folch, a biologist working for the Aragon regional government, had taken skin scrapings from Celia's ears and stored the tissue samples in liquid nitrogen in order to preserve the bucardo's genetic line. A few years later, in 2003, Folch and his group removed the nucleus from one of Celia's ear cells, transferred it into an egg cell of a domestic goat, and implanted it into a surrogate mother in a procedure called interspecies nuclear transfer cloning.

After a gestation period of five months, the surrogate mother gave birth to a live Pyrenean ibex. By any standard, this was an astonishing event. After being officially, literally, and totally extinct for more than two years, a new example of the vanished species was suddenly alive and breathing.

Not for long, however. The baby ibex lived for only a few minutes before dying of a lung condition. Still, those scant minutes of life were proof positive that an extinct species could be resurrected, not by magic or miracles but by science.

"Nuclear DNA confirmed that the clone was genetically identical to the bucardo's donor cells," the group wrote in its report on the project. "To our knowledge, this is the first animal born from an extinct subspecies."

Almost certainly, it will not be the last. The bucardo's birth involved a bit of genomic reprogramming because the egg cell that developed into the baby ibex had not been fertilized by a sperm cell but rather by the nucleus of a somatic (body) cell. The nucleus and the egg cell had to be jump-started into becoming an embryo in a process known as electrofusion, which melds the two together.

A later technique under development in my Harvard lab will allow us to resurrect practically any extinct animal whose genome is known or can be reconstructed from fossil remains, up to and including the woolly mammoth, the passenger pigeon, and even Neanderthal man. One of the obstacles to resurrecting those and other long extinct species is that intact cell nuclei of these animals no longer exist, which means that there is no nucleus available for nuclear transfer cloning. Nevertheless, the genome sequences of both the wooly mammoth and Neanderthal man have been substantially reconstructed; the genetic information that defines those animals exists, is known, and is stored in computer databases. The problem is to convert that information—those abstract sequences of letters—into actual strings of nucleotides that constitute the genes and genomes of the animals in question.

This could be done by means of MAGE technology—multiplex automated genome engineering. MAGE is sort of a mass-scale, accelerated version of genetic engineering. Whereas genetic engineering works by making genetic changes manually on a few nucleotides at a time, MAGE introduces them on a wholesale basis in automated fashion. It would allow researchers to start with an intact genome of one animal and, by making the necessary changes, convert it into a functional genome of another animal entirely.

You could start, for example, with an elephant's genome and change it into a mammoth's. First you would break up the elephant genome into about 30,000 chunks, each about 100,000 DNA units in length. Then, by using the mammoth's reconstructed genome sequence as a template, you would selectively introduce the molecular changes necessary to make the elephant genome look like that of the mammoth. All of the revised chunks would then be reassembled to constitute a newly engineered mammoth genome, and the animal itself would then be cloned into existence by conventional interspecies nuclear transfer cloning (or perhaps by another method, the blastocyst injection of whole cells).

The same technique would work for the Neanderthal, except that you'd start with a stem cell genome from a human adult and gradually reverse-engineer it into the Neanderthal genome or a reasonably close equivalent. These stem cells can produce tissues and organs. If society becomes comfortable with cloning and sees value in true human diversity, then the whole Neanderthal creature itself could be cloned by a surrogate mother chimp—or by an extremely adventurous female human.

ఇ ఇ ఇ

Any technology that can accomplish such feats—taking us back into a primeval era when mammoths and Neanderthals roamed the earth—is one of unprecedented power. Genomic technologies will permit us to re-play scenes from our evolutionary past and take evolution to places where it has never gone, and where it would probably never go if left to its own devices.

Today we are at the point in science and technology where we humans can reduplicate and then improve what nature has already accomplished. We too can turn the inorganic into the organic. We too can read and in-terpret genomes—as well as modify them. And we too can create genetic diversity, adding to the considerable sum of it that nature has already produced.

In 1903 German naturalist Ernst Haeckel stated the pithy dictum "On-togeny recapitulates phylogeny." By this he meant that the development of an individual organism (ontogeny) goes through the major evolutionary stages of its ancestors (phylogeny). He based this aphorism on observa-tions that the early embryos of different animals resembled each other and that, as they grew, each one seemed to pass through, or recapitulate, the evolutionary history of its species. (For example, the human embryo at one point has gill slits, thus replicating an evolutionary stage of our piscine past.)

While it is clear that embryos develop primitive characteristics that are subsequently lost in adults, Haeckel's so-called biogenetic law is an over-statement and was not universally true when first proposed or today. How-ever, I hereby propose a biogenetic law of my own, one that describes the current situation in molecular engineering and biotechnology: "Engineer-ing recapitulates evolution." Through human ingenuity, and by using the knowledge of physics and chemistry gained over the course of six indus-trial revolutions, we have developed the ability to manipulate and engineer matter, and by doing so we have rediscovered and harnessed the results of six similar revolutions that occurred during billions of years of biolog-ical evolution.

Using nanobiotechnology, we stand at the door of manipulating genomes in a way that reflects the progress of evolutionary history: start-ing with the simplest organisms and ending, most portentously, by being able to alter our own genetic makeup. Synthetic genomics has the potential to recapitulate the course of natural genomic evolution, with the difference that the course of synthetic genomics will be under our own conscious deliberation and control instead of being directed by the blind and opportunistic processes of natural selection.

We are already remaking ourselves and our world, retracing the steps of the original synthesis—redesigning, recoding, and reinventing nature itself in the process.

CHAPTER 1

-3,800 Myr, Late Hadean
At the Inorganic/Organic Interface

What follows is the greatest story ever.

It's the story of a once invisible being, nameless for eons, now called "the genome." Its being—its existence across time, its depth and complexity as a natural artifact, and the vast abundance and variety of its manifestations—is the story. It is ancient and modern, older than our oldest ancestor and yet fresher than a newborn baby. It has covered our planet with its descendants, now over a billion times a billion times a billion copies (10^{27}).

The tale of the genome involves more sex than the most pornographic novel imaginable. The narrative is replete with incredible action scenes, countless life-and-death struggles, wild improbabilities that turn out to be true, and overwhelming successes in the face of staggering odds. It is a story about families and universal truths. In the retelling, it becomes, in part, your own personal story. The tale reveals a vibrant past and may lead us to a better future. As the ultimate self-help manual, it offers better health and longer life, along with "descendants as numerous as the stars in the sky and as the sand on the seashore" (as in the Judeo-Christian-Islamic tradition), or "as numerous as the sands on the Ganges" (in Buddhism).

As befits the greatest story ever, this is a multiplex tale, enacted and told in a spiral of understanding. Through its abundance, fidelity, and diversity, the genome adapted to the physical world, solving a small number of basic problems repeatedly, passing on the answers, and occasionally even rediscovering solutions once lost. We see these problems solved in the first instance biologically, by the process of evolution. Nature turned inorganic materials into organic substances. Natural organisms read and interpreted genomes. And natural organisms have created huge amounts of genetic diversity. That network of natural interactions comprises our first tale.

It begins long ago, in the Hadean era.

Can Organic Arise from Inorganic? Selection Among Atoms and Molecules

The Hadean geologic era lived up to the image of an underworld inhabited by the ancient Greek god Hades—lifeless and full of hot lava—3.8 billion years ago. If a living cell were unfortunate enough to travel back through time to the Hadean landscape, it would be cooked: all water vaporized and its precious complexity of living stuff dry-roasted and then mineralized, turned from delicate, filmy proteins into charcoal (graphite), water vapor, and other waste products.

Before this, all the way back to the big bang, the universe was made up almost entirely of hydrogen nuclei, the simplest of all elements, consisting of just one proton. These protons would collide and fuse together to form helium nuclei (2 protons). Inside stars these helium nuclei would in turn fuse to form carbon (6 protons). Carbon nuclei would then enter a cycle (the carbon-nitrogen cycle), taking in hydrogen, and by adding nitrogen (7) and oxygen (8) intermediates, would catalyze the formation of yet more helium. The new helium would, as before, make more carbon. The net outcome of all this is that in hot stars carbon catalyzes the formation of copies of itself. (By "catalyze," I mean causing or accelerating a reaction without the catalyst itself undergoing a permanent change.)

These thermonuclear transformations, which occur at Hades-plus temperatures within stars, are accompanied by the release of enormous amounts of energy in the form of radioactive particles such as gamma ray photons, positrons, and neutrinos. (And also of course by the heat and light that drive life on this planet.)

The processes that make up the carbon-nitrogen cycle can be thought of as a form of natural selection for favorable reactions and stable elemental forms (atoms and their isotopes). This seems analogous to the mutation and selection of living species, and still later the mutation and selection of synthetic organisms. Today those five (hydrogen, helium, carbon, nitrogen, and oxygen) of the eighty stable elements are the most abundant in the universe. These processes selectively skipped over weakly represented lithium (3), beryllium (4), and boron (5).

A list of such atomic elements (substances that chemically cannot be broken down further) is a prerequisite for understanding the next level of selection complexity—the combination of those basic atoms into the compounds (molecules) of nature. Antoine Lavoisier wrote the first comprehensive list of the elements in the first modern chemistry text, *Traité élémentaire de chimie*, in 1789. He listed thirty-one in all, together with light and "caloric" (heat), making up a total of thirty-three "simple substances belonging to all the kingdoms of nature, which may be considered the elements of bodies." As Lavoisier presented them:

LIGHT	SULFUR (S)	ANTIMONY (SB)	MERCURY (HG)	CALCIUM (CA)
CALORIC	PHOSPHORUS (P)	ARSENIC (AS)	MOLYBDENUM (MO)	MAGNESIUM (MG)
OXYGEN (O)	CARBON (C)	BISMUTH (BI)	NICKEL (NI)	BARIUM (BA)
NITROGEN (N)	CHLORINE (CL)	COBALT (CO)	PLATINUM (PT)	ALUMINUM (AL)
HYDROGEN (H)	FLUORINE (F)	COPPER (CU)	SILVER (AG)	SILICON (SI)
	BORON (B)	GOLD (AU)	TIN (SN)	
		IRON (FE)	TUNGSTEN (W)	
		LEAD (PB)	ZINC (ZN)	
		MANGANESE (MN)		

Each element in the table above is followed by the abbreviation that is commonly used in most branches of science, and even within the general culture—for example, H_2O (water), NaCl (salt), and CO_2 (carbon dioxide).

Jöns Jakob Berzelius, who developed an interest in chemistry in medical school, introduced these symbols in 1813. By 1818 he had measured the masses of forty-five of the eighty stable elements. As we will see in Chapter 3, as few as six elements may be sufficient to create the major molecules of life: S, C, H, P, O, N (sulfur, carbon, hydrogen, phosphorus, oxygen, and nitrogen—pronounced "spawn"—shaded gray in the table above). These constitute the most abundant elements in living systems; also needed are metal ions such as magnesium (Mg) that are involved in key reactions of these compounds.

These elements chemically combined with one another to form molecules, such as water, as the newly formed earth cooled. How did life arise from nonlife? To understand this, we need to explore the universe of simple, nonliving chemicals. As far as we know, the physical and chemical properties of the elements are set largely by particles in the nucleus (as well as by those in the surrounding electron cloud), and not by the specific arrangement of those particles. For example, it matters only that there are six protons in carbon; the exact structural relationships among the protons are irrelevant. Those six protons, irrespective of how they are arranged in the nucleus, attract and retain an equivalent number of electrons in the surrounding electron cloud.

In molecules, by contrast, the physical arrangement of the component atoms is crucial. For example, a molecule of water, H_2O, is not just ten protons and ten electrons packed together randomly in a jumble. The order of the atoms and their shape matters. Water is not H-H-O but rather H-O-H, meaning that each hydrogen atom can only bind to the oxygen atom, and not to two atoms. Molecules are like intimate social networks. Some atoms, such as hydrogen, tend to make single bonds with only one other atom. Oxygen makes two bonds, nitrogen three, while an atom of carbon can bond with four other atoms. So, water has each hydrogen bonding with one atom, oxygen, and its oxygen bonding with two atoms.

Let's now replace each hydrogen in water with a carbon (keeping each carbon happy with its own three hydrogens): this will give us dimethyl ether, CH_3-O-CH_3. So let's check the bonds. The oxygen still has two single bonds—one to each carbon—and each carbon has four single bonds, three to hydrogens and one to the central oxygen.

$$
\begin{array}{ccccc}
& H & & H & \\
& | & & | & \\
H - & C & - O - & C & - H \\
& | & & | & \\
& H & & H &
\end{array}
$$

Now we can illustrate the importance of spatial arrangement. If we keep all nine component atoms but rearrange them slightly, say to CH_3CH_2OH, we get a radically different set of physical and chemical properties in a molecule called ethanol.

$$
\begin{array}{cccc}
& H & H & \\
& | & | & \\
H - & C & C & - O - H \\
& | & | & \\
& H & H &
\end{array}
$$

What a difference that simple rearrangement makes! Dimethyl ether boils at -24 degrees C while ethanol boils at +78 degrees C. Many people like to drink ethanol (typically 8 to 15 percent in water), but you would not want to drink dimethyl ether. These rearranged molecules are called isomers of each other (Greek for "the same parts"). Ethanol is an isomer of dimethyl ether: each molecule has two carbons, six hydrogens, and one oxygen, but differently arranged.

Berzelius came up with the concepts and terms for catalysis, polymer, and isomer, among others. He also provided experimental evidence for the law of definite proportions (first stated by the French chemist Joseph Proust), which holds that the proportions of the elements in a compound are always the same, no matter how the compound is made. Even though we have been introducing these ideas by appealing to the simple bonding of discrete atoms, Berzelius discovered them by doing two thousand analyses over the course of a decade, purifying and weighing chemicals and their reaction products. He noticed that the ratios were reproducible and generally came in values that were expressible in whole integers. Berzelius was also the first to recognize the difference between organic compounds that were derived only from living matter, and all other chemicals, which he lumped together as "inorganic." This distinction contributed greatly to our understanding of life and set the stage for inquiries into vitalism, the theory that life and its processes are not reducible to the laws of physics and chemistry. Berzelius believed that something kept living matter distinct from nonliving matter. But work done in four areas—the synthesis of urea, the investigation of mirror molecules, the investigation of polymers (especially of the DNA/RNA polymers), and the self-reproduction of molecules—argues to the contrary.

Berzelius's protégé Friedrich Wöhler also came to chemistry through the study of medicine. In 1828 Wöhler (accidentally) became the first person to synthesize an organic compound, urea, from an inorganic substance, ammonium cyanate. The reaction in question is $NH_3HNCO \rightarrow NH_2CONH_2$. This is a rearrangement of atoms similar to that of the isomers mentioned above. But at the time it was more mysterious, in part because the description of chemicals as precise arrangements of atoms was just becoming evident from experiments. Second, urea was thought to come only from the urine of certain vertebrates as well as, less obviously at the time, other species. Ammonium and cyanate were considered to be inorganic components of minerals.

Wöhler's synthesis of urea was arguably the first great challenge to vitalism. Since then, scientists have tried to make ever more complex organic living systems from inorganic or otherwise simple nonliving atoms and molecules. With hindsight, urea was a very simple case (consisting of just eight atoms of carbon, hydrogen, oxygen, and nitrogen) and was thus poised for success in this first of five grand challenges to vitalism—all of which reflect milestones in practical synthetic biology as well.

The second challenge to vitalism concerns the phenomenon of the handedness of molecules—one of the distinguishing features of living systems. The challenge is to determine whether natural single-handedness can arise spontaneously or be reversed, and if so, what the consequences would be.

The chemistry of life is based on polymers made by linking monomer molecules together in long linear sequences, just as written texts are made of linear sequences of letters. These two terms share the common root "mer," from the ancient Greek *meros* for "part." A monomer, accordingly, is a single molecule (one part), whereas a polymer (many parts) is a molecular structure composed of many similar molecular units bonded together. Amino acids are monomers whereas combinations of them are polypeptides (a.k.a. proteins), which are polymers. The large molecules known as RNA and DNA are also polymers—polynucleotides—consisting of many simple molecular subunits known as nucleotides. Those three types of polymers can bind and catalyze the formation of other polymers as well as the metabolism of the basic components of living things. A single typo in a biopolymer sequence could make the polymer nonfunctional and nonliving. So the *third challenge to vitalism is to find out whether those long, precise sequences could arise spontaneously and possess the functions of life such as catalysis.* Can new kinds of life exist that have no ties to ancient life—a truly artificial or synthetic life form?

The fourth challenge is determining whether a fully synthetic chemical network could make a copy of itself and evolve (i.e., change with time) and in so doing, prolong its own survival. And *the fifth challenge is whether consciousness (or a mind) can arise synthetically.* This will be addressed in the Epilogue.

Is Biological Handedness Special? What Are the Consequences of Reversing It?

This section will consider the second challenge to vitalism: biomolecular handedness. There are six compelling reasons to care about handedness.

First, when we inspect meteorites and other matter that has fallen to the earth from space, we look for an excess of molecules of the same handedness (one "enantiomer," meaning one of a pair of molecules that are mirror images of each other). In space there are more molecules of one specific handedness than of the other. Does this mean that life arose far away and landed here, or rather that one hand is more likely to spontaneously arise or survive? The answer to this question has profound implications for our place in the universe.

Second, the two different hands have different pharmacological effects. The drug thalidomide was used in Europe between 1957 and 1961 to treat morning sickness in pregnant women. Thalidomide was made chemically and not biologically and hence both hands were made in relatively equal amounts. It turns out that one hand cures the morning sickness while the other causes severe limb malformations in the developing fetus (a result described by the BBC as "one of the biggest medical tragedies of modern times").

Third, chemicals whose molecules exist in only one spatial arrangement tend to be more economically valuable than those that are mixtures of molecules having a given arrangement together with those of their mirror images. The "unnatural" versions are more expensive (1,400-fold more for the amino acid isoleucine).

Fourth, the oceans contain a large mass of carbon trapped in the form of recalcitrant dissolved organic matter (the ominous sounding RDOM), much of which consists of mirror-image forms of easily recycled (nonrecalcitrant) matter. The handedness of these trapped carbon molecules causes them to persist in the oceans for millennia.

Fifth, the ability to reverse the handedness of useful polymers, such as cellulose, wool, and silk, could retard decay. Biodegradable plastics may come to be seen as a mixed blessing. The usual route of biodegradation is through release of carbon dioxide, which is currently an unwelcome output. Also, the energy normally expended in recycling or replacing degraded polymer products might be saved in some cases.

Sixth, at the extreme, a mirror cell or a mirror organism (composed of chemicals of reversed handedness) might be resistant to all or nearly all parasites and predators, a tremendously valuable result.

Since biomolecular handedness is so important, what is it? The basic idea is conveyed by the fact that our right and left hands are mirror images of each other and are not related by simple rotations. If we take a sculpture of a right hand and press it into a soft mold, we will discover that we cannot fit our left hand into the mold (Figure 1.1). However, if we fill that mold with plaster, the resulting new copies are considered complementary and are of the same handedness as the originals.

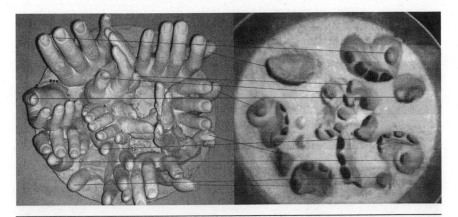

Figure 1.1 A sculpture on the left and corresponding negative mold—illustrating handedness and complementary shapes.

This same phenomenon exists on the molecular level. For example, there are two ways to arrange the four atoms that can bond to a carbon atom, and each will be a mirror image of the other. Furthermore, each will have predictably similar properties.

This left-right feature is also known as chirality, from the Greek (χείρ) for "hand." Even scientists who don't think about mirror worlds initially show great confusion as to whether the properties of mirror versions of molecules, cells, and bodies can be accurately predicted based on the properties of their nonmirror versions. Consider this. If you build a replica of an old-fashioned clock by only looking at its reflection, the copy will predictably tell time, but the numerals will be mirror images of the originals and the hands will rotate counterclockwise. These outcomes are precisely as anticipated.

Here's a simple demonstration that relates the hands and clock examples to molecules. Start with a central cantaloupe ball, and use toothpicks to successively place around it, in a clockwise order, a raisin, a piece of coconut, and a piece of nectarine all flat on the table. Then make another such structure using the same pieces of fruit but placing them counterclockwise. You can flip one over so that the two structures match, but if you add a bit of honeydew above and attached to the central cantaloupe, then no matter how you orient the structures you can still tell which was

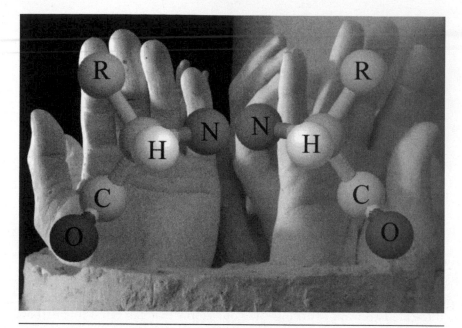

Figure 1.2 Another example of handedness uses a ball and stick model of a carbon atom with its four bonds (to H, CO, R, N groups).

clockwise originally and which wasn't. If you place them in front of a mirror you can see that they are each other's mirror image. Now let's replace fruit with atoms: H, CO, R, N. This is the general structure of an amino acid. The NH_2 is the amine and COOH is the acid. R refers to a "radical" (a group of atoms that behave as a unit) that varies with amino acid type.

Amino acids have a known handedness. You can impress your friends by your ability to identify the natural form. In nature, for reasons still unknown, almost all biomolecules vastly prefer one of the two hands (amino acids and proteins being designated as left-handed). Life itself, in a way, is fundamentally single-handed. Here is a procedure for telling whether a human hand, or a molecule, is right- or left-handed. Looking at your left hand palm up as in Figure 1.2, go from thumb to index finger to pinkie, the direction is clockwise, which indicates left-handedness. Performing the same observation on the right hand gives a counterclockwise direction, indicating right-handedness.

Now let's do the same for molecules. When looking down the bond from the hydrogen (H) to the central carbon, if the other groups going clockwise are CO, R, N, as on the left of Figure 1.2, then the configuration is normally seen in natural proteins (sometime called *levo* or L, for left-handed). On the right is the mirror version (*dexter*, Latin for "right," or D). The R (radical) group distinguishes the twenty (or so) types of amino acids, each with its own personality (and its own single-letter code). Some are electrically negative while others are positive. Some are greasy and fear water (or hydrophobic), while others love water (hydrophilic). Glycine is the only amino acid that is its own mirror image, since its two hydrogen atoms are normally indistinguishable. Just to keep us on our toes, natural nucleic acids (RNA and DNA) were long ago designated D and their mirror forms L.

By now you may be wondering about the cash value of this talk about handedness. Just as there can be mirror molecules, there can also be mirror life. Mirror life would be the result of changing the handedness of an entire organism and all of its components, so that you have a mirror image of everything from the macro level all the way down to the atomic level. While mirror life may *look* identical to current life, it would be radically different in terms of its resistance to natural viruses and other pathogens. Mirror life forms would be immune to viruses and other pathogens, the reason being that the molecular interactions of life are exquisitely sensitive to the mirror arrangement of their component atoms and molecules. Normal viruses would not recognize a mirror organism as a genuine life form whose cells it could invade and infect. Such multivirus resistance would be an incredible boon to humanity. But it would come at a steep price because mirror life would be unable to digest foods by means of normal enzymes, which would mean that we would need to develop, cultivate, and mass-produce a whole range of mirror foodstuffs. (Although biohackers could in principle synthesize mirror viruses and other pathogens, mirror humans would still be resistant to natural pathogens—and even to genetically engineered nonmirror superpathogens.)

The prospect of mirror humans raises unusual and startling possibilities. Supposing that there will be a transition to a mirror version of human

beings at some point in the distant future, the changeover would be grad-
ual, with a substantial interregnum period when two types of human beings would exist: natural humans composed of natural-handed molecules,
and mirror humans made up of mirror versions of them. In this situation,
it's almost as if two separate species of humans existed simultaneously. Or
we might see an equilibrium between the two types if mirror pathogens
arose.

These mirror humans should have an unusual smell. Members of the
two versions could marry, but producing children would require what
today would be considered extraordinary efforts. However, by the time we
can make mirror humans, making designer (or random) children of either
mirror type will not seem as challenging as it does today. We might even
be able to make mirror identical twins or bodies that are mixtures of both
types of cells.

Finally, creating a race of mirror humans is not without risks. Although
new mirror molecules interact with mirror versions of existing molecules
in predictable ways, how they interact with biomolecules in general is unpredictable. However, they are no more unpredictable than any newly synthesized drug, chemical, or material; nevertheless, careful screening of
mirror molecules by computational methods or by actual experiment will
be necessary to ensure safety.

ॐ   ॐ   ॐ

Louis Pasteur had the first inklings into what natural chemical chirality is
all about. He acquired this understanding by performing what the magazine *Chemical and Engineering News* once referred to as the "most beautiful chemistry experiment in history."

Pasteur's seventy-two years on earth are remembered mainly for his
contributions to microbiology—especially for inventing pasteurization,
discovering the "Pasteur effect" (the anaerobic growth of organisms), developing the first vaccines for rabies and anthrax, and contributing to the
understanding of fermentation, as well as for his clever experiments in
support of the germ theory of disease. Nevertheless, his earliest and

equally great achievements came from his work as a crystallographer. In 1848, the seventy-three-year-old French physicist Jean Baptiste Biot sponsored the twenty-five-year-old Louis Pasteur in his first experiments at the elite college École Normale Supérieure in Paris.

This is a story about tartar, a chemical extracted from grapes. The modern chemical term "racemic" (meaning a mixture of the two hands) comes from the Latin *racemus* for "cluster of grapes." In 1838 Biot found that tartaric acid, unlike its isomer, racemic acid, was optically active, meaning that it rotated the plane of a beam of polarized light. Both isomers are found in wine, the latter in sediments or by heating tartaric acid. These two acids were one of the first examples of an isomer pair and they turned out to be unusual in that almost all of their physical and chemical properties were identical except for their solubility and their ability to rotate polarized light.

Pasteur first showed that an equal mixture of the two forms of a salt of tartaric acid will spontaneously separate into small crystals. He heroically separated these microscopic crystals with tweezers and then dissolved them and showed that the solutions rotated polarized light in opposite directions. This key optical property is independent of orientations of the molecules, since in solution the molecules can take on all possible orientations. Furthermore, he made the profound observation that tartaric acid originating from natural systems (yeast from his wine making) came in only one of the two-handed states, the left-handed version. Since then we have learned that this propensity for single-handedness is a characteristic of most molecules that are constituents of living systems.

Omne vivum ex vivo: "all life from life," the irreducibility of living matter to anything nonliving, became one of Pasteur's most strongly held convictions. Indeed, Pasteur's key role in debunking the theory of "spontaneous generation of life" in 1859 could be construed as equally aimed at propping up the idea that the living forms of chemicals cannot be synthesized. That idea was consistently opposed by French chemist Marcelin Berthelot, who believed in the power of synthetic chemistry and experimentally synthesized organic substances that did not occur in nature. The dispute between Pasteur and Berthelot was not settled until 1971, when

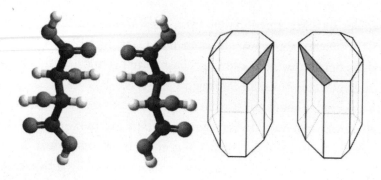

Figure 1.3 The mirror forms of tartaric acid at the atomic level on the left and the corresponding crystals of the salts: for each pair the natural one is on the left. The handedness at the atomic (nanometer) scale building blocks is amplified up to the human (millimeter) scale, a key requirement at the time of Pasteur, since the ability to manipulate molecules was still quite primitive. Just as the arrangement of the atoms of the two molecules can only be superimposed by use of mirrors, not by mere rotations, the same is true of the arrangement of the faces of the two crystals.

H. P. Kagan synthesized the aromatic chemical helicene, a helical-shaped molecule, following Le Bel's 1874 proposed asymmetric synthesis using circularly polarized light.

The future of chirality is bright with promise and includes the synthesis of mirror cells that will give us access to valuable chemicals and materials, as well as cells that resist most biological degradation. Heat can often reverse the handedness of a given molecule, and a class of enzymes called racemases can do the same. The most obvious way of creating a mirror cell would be to synthesize unnatural mirror versions of all living molecules and then assemble the parts in the correct spatial arrangements into a full cell and hope that it replicates.

The enormity of such a task will become more evident in Chapter 2, where I describe modern cells and minimal versions of cells. I show a possible shortcut—that by leveraging a slightly sloppier version of current life we could avoid the precise manual synthesis and assembly of billions of chemical bonds. But for now we want to consider the implications, regardless of route. The complete synthesis of a mirror life form from the atoms up would be the next, and perhaps final, step in the overthrow of vitalism. The shortcut version would pose less of a threat to vitalism but would be just as significant in terms of its potential future applications.

Creating a mirror world might give us a fresh lease on life, one free of disease and unwanted agricultural pest species, but subject to unintended consequences such as turning too much carbon dioxide into RDOM or encouraging the proliferation of enzymes that could attack our wonderful new mirror life—enzymes that are currently rare but in a sense are waiting for a justification for their duplication, diversification, and optimization. Unlike antibiotics, of which there are thousands of natural and totally synthetic examples to draw upon, there are only two hands, and we've already "used up" one of them.

Another profound implication is that we are doubling the number of chiral chemicals in our bag of tricks. Because many chemicals have more than one chiral atom, the number of new compounds might be exponential—as high as 2^N where there are N atoms with the property of mirror asymmetry. This could be especially apparent in polymers—the topic of the next section.

Can a Synthetic Chemical Copy Itself and Evolve Without Help from Living Systems?

Having journeyed from inorganic to organic and having considered the handedness of simple monomers, we now take a look at polymers, the next big idea in the story of the past and future of life. The key substances in most modern biological structures and catalysts are proteins, which are polymers made up of the twenty or so amino acids. The other key classes of polymers are the polynucleotides DNA and RNA, each of which is composed of four nucleotides (adenine, cytosine, guanine, and thymine in DNA, with uracil replacing thymine in RNA). Like proteins, DNA and RNA have handedness; Tom Schneider (a molecular-machine researcher at the US National Cancer Institute) maintains a web page (http://www.fred.net/tds/leftdna) dedicated to seven hundred humorous examples of wrong-handed DNA helix art in the popular press and in company brochures. (We need to create mirror DNA to make these folks seem more prescient and less foolish!) Like polypeptides, polynucleotides are capable of structural scaffolding and catalysis but have the additional feature of making replication possible (indeed obvious and compelling).

The story of DNA (and RNA) replication is beautifully simple. The core idea is that of complementary surfaces. Just as the pairs of hands and molds in the sculpture shown in Figure 1.1 are uniquely complementary, so too are the base pairs of RNA: U (uracil) bonds with A (adenine), and C (cytosine) with G (guanine), Figure 1.4. The idea that a DNA duplex explains DNA duplication, and these two base pairs in particular, won Jim Watson and Francis Crick the Nobel Prize. By contrast, the other eight possible pairings (AA, AC, AG, CC, CU, GG, GU, and UU) are much weaker—energetically close to no pairing at all. In the case of the strong pairings, the two surfaces fit together, so to speak, whereas in the case of the weak pairings, there is a mismatch between the respective molecular shapes.

The stability of two strands of RNA binding to each other depends essentially on their length. Given a long polymer of nucleotides, we can assemble matching monomers, or short complementary polymers, onto it by base pairing. The stability of the resulting double helices gives the second strand of RNA a chance to polymerize or ligate (join) the short bits into a new long polymer (probably catalyzed by molecules floating about). This new polymer is not identical to the original but complementary, but the complement of the complement is the original.

Earlier I asked how we can get from atoms to replicable structures from scratch, meaning from atoms or tiny clusters of atoms without assistance from living templates. We can now see a progression of events from the beginning of this chapter as atomic nuclei randomly join to become atoms, atoms join to become molecules, and monomeric molecules randomly join up to become polymers. The chemical bonding ratios up to this point seem predetermined by physical selection rules acting on large sets of atoms. The matter that we see in the cosmos requires for its existence only a one part per billion excess of matter over antimatter in the early universe.

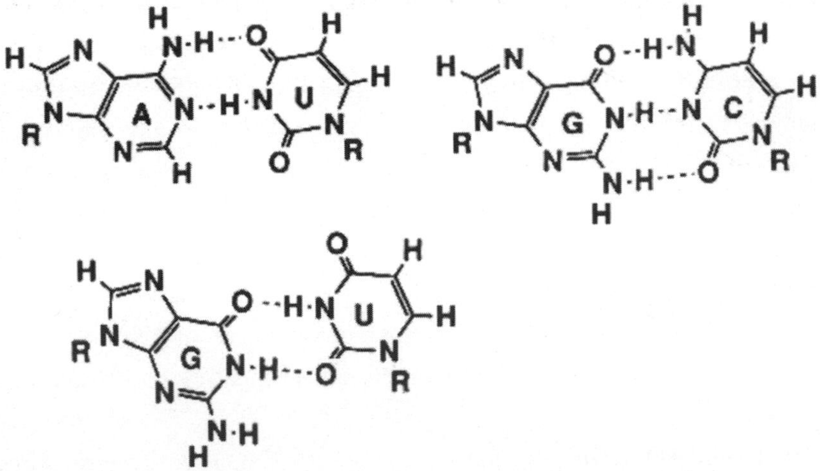

Figure 1.4 Complementary shapes of RNA base pairs. Shown at the top are the two dominant base pairs (AU and GC); just below them is an example (GU) of the other eight possible (very weak) base pairs. What seems like a subtle difference in geometry between the AU pair and the GU mispair makes for a huge difference in the context of a stack of these flat base pairs in the double helix. (The R represents the ribose sugar to which all four bases bind with very similar geometry.) The dotted lines are bonds mediated by hydrogen that are about one hundred times weaker than the covalent bonds (solid lines) in their optimal configuration.

Had there been exactly equal amounts, they would have annihilated one another. There is no consensus yet on the explanation of this asymmetry. Similarly, if there had been equal amounts of left- and right-handed molecules, life might not exist in the universe—at least not life as we know it. In any case, once we get replication, then we can expect to see, more and more frequently, small random events that can grow exponentially into interesting structures before any competing chemistry can take hold.

Evolution happens not only in nature but also in the laboratory, where the key processes of mutation and selection operate on inanimate molecules and structures made up of them. Even creationists can see how small changes, when made repeatedly over long stretches of time, can add up to enormous effects that confer substantial selective advantages on a given organism. What is more remarkable is how new kinds of functionality and shape can emerge out of totally random collections of RNA rather than as mere variations on something already optimized and working. This process of emergence has major implications for how quickly new genes and genomes could have arisen in the past, as well as for the design of medical and industrial materials in the near future. Totally random libraries of RNA can be subjected to powerful selection pressures that favor rare molecules capable of valuable binding or catalysis functions. We can generate an incredible number of different RNA structures in a volume equivalent to that of a small cell. If any of these RNAs has any activity for preferentially cutting and/or joining, then the whole set of RNA sequences could churn and self-modify until stable self-replicating molecules arise and persist.

So, the answer to the question posed earlier—Can a synthetic chemical copy itself and evolve without help from living systems?—is a resounding yes. Here is an example of such evolution in the lab. A molecule of theophylline (which is used as a drug to treat asthma and other lung diseases) can form part of a fifty-five-nucleotide-long stretch of RNA that can have two different morphologies and two different functional states depending on the concentration of theophylline. It is easy to imagine that this molecule could start with either state as its "only" shape and function and could change to the bi-stable shape with as little as the mutation of a single nucleotide. Then after some other molecule adapts to the bi-stable state, another point mutation locks it into one state or the other, permanently.

The moral of the story is that shape and function can be altered radically with just a few changes that nevertheless yield a selective advantage at each separate stage. This capacity will be very handy in the future of lab-evolved designs.

The Future Interface of Inorganic and Organic Worlds

We have been focusing on inorganic and organic chemistry. In colloquial usage the term "organic" is attended by a certain halo effect that, upon analysis, it doesn't deserve. When we buy organic produce, we are supporting the idea of feeding crops the essential elements nitrogen and phosphorus that are derived only from animal excrement rather than from conventional mineral fertilizers like ammonium phosphate as churned out by the chemical industry. Does this sound like a latter-day vestige of vitalism? These organic fertilizers obviously bear a public health risk in the form of fecal pathogens such as *E. coli* 0157:H7, *Cryptosporidium,* and *Giardia*. Both methods of fertilization, if used to excess or done poorly, carry a risk of run-off into streams and ponds resulting in fish kills.

Another inorganic/organic dualism can be seen at the interface between life and machines. I/O means not only the intimate dance of inorganic/organic, but also input/output. Today scientists are recapitulating what we might call the first inorganic/organic transition that occurred eons ago. We take simple molecules and form them into linear polymers that are the building blocks of both natural and synthetic structures. We increasingly want to see input/output between inorganic electronics and organic DNA. On the input side of I/O, megapixel CCD (charge-coupled device) and CMOS (complementary metal oxide semiconductor) electronic cameras can be used to record spatially patterned light, such as bioluminescence or fluorescence, to inorganic (i.e., silicon-based) computers. This would allow us to read genomes speedily, whether for diagnostic testing or environmental monitoring. Coupling these inorganic/organic, input/output features together permits us to design, synthesize, and assess the quality of large collections of DNA and anything that they encode.

Back in the early stone age of DNA engineering (circa 1967–1990) we made DNA in solution and had to purify very short intermediate products. The low yields for each step, multiplied by the short lengths per step, made DNA synthesis a challenging, tedious enterprise. Nowadays we can literally "print" arrays of DNA by machine. This is a really big deal. To see why, let's explore analogies with other types of printing.

Today we use spatially patterned light and optics or ink-jet printers to print photographs on paper, which are two-dimensional artifacts. But it is possible for those same ink-jet printers to "print" (i.e., to construct, layer by layer) three-dimensional objects. Ink-jet systems can hold many colors and activate many jets in parallel. If the ink consists of colored minerals or glue, then we can deposit (or "print") one layer on top of a second layer (typically 0.1 mm per layer), and then repeat this process successively to create three-dimensional rapid prototypes of artifacts in plastic or plaster.

We can use similar approaches of spatially patterned light or ink jets to build up long chains of DNA called oligonucleotides, or "oligos" (from the Greek *oligos*, for small), up to 300 nucleotides in length. Typically each layer is one nucleotide (= 0.4 nm) thick, with the four ink-jet "colors" (A, C, G, and T) used per layer. By this method we can make millions of different patches of DNA on a 3- by 1-inch glass slide or portion of a larger silicon wafer.

In 1980 commercial DNA synthesis services were available, at the going rate of $6,000 for a small amount of product, only about ten nucleotides long. They were used either to find valuable genes in cellular RNA or to synthesize them. By 2010 we could make a million 60-nucleotide oligos for $500. Just as the global appetite for reading DNA seems insatiable—growing a million-fold in six years and still increasing—the appetite for DNA synthesis, or "writing," will probably grow similarly and go in many unexpected directions. Since DNA in cells is very long-lived (billions of years), we might want to preserve the whole Internet in the form of DNA molecules. This would be the ultimate backup, made possible by converting the Internet's 0s and 1s to the DNA molecule's As, Cs, Gs, and Ts, and synthesizing the molecules accordingly. The Internet Archive contains 3 petabytes (10^{15}) of data, and is expanding at the rate of 1 petabyte per year. This granddaddy of all backup copies would cost $25 billion, an amount that is not out of the question, but bringing that cost down by three to six factors of ten would be desirable. Because of its very small size, launching copies into space and icy moon polar craters could be very inexpensive.

Today, oligonucleotide chips are becoming the lifeblood of synthetic biology. However, spatially patterned light and ink-jet printers can be used to make objects as complex as patterned cells. Various options exist: (1) the cells themselves can be shot directly from ink jets, (2) scaffolding proteins can be deposited in such a manner that the cells self-assemble onto those proteins, or (3) the cells can be assembled onto photo-reactive scaffolding and then selectively stabilized or released by light. These and other methods hold the potential of making synthetic and even personalized tissues and organs suitable for testing pharmaceuticals—and ultimately for printing copies of whole organisms.

As we go forward we will be seeing more hybrid inorganic/organic systems. Our children already inherit our mechanically augmented biology, in the form of cars, smart-phones, hearing aids, pacemakers, and so on, and these devices have become increasingly integrated into our daily lives; indeed, many people would find it hard to live without them. Since the 1980s we have added recombinant DNA-based parts to our bodies in the form of insulin, erythropoietin, monoclonal antibodies, and other medically useful substances. The addition of complex synthetic biological systems to this mix will ultimately blur the distinction between life and nonlife.

CHAPTER 2

-3,500 Myr, Archean

*Reading the Most Ancient Texts
and the Future of Living Software*

ॐ

The greatest story ever—the story of the genome—continues through the Archean geologic era, which started roughly 3,500 million years ago. The name "Archean" stems from the Greek *arche*, which means "the beginning." (The same ancient Greek term also refers to the keel of a boat, the part from which everything else rises.)

Back in the Archean, the earth scarcely resembled the planet that exists today. For one thing, there was no free oxygen in the atmosphere, which consisted largely of gases such as methane, ammonia, hydrogen sulfide, and the like, a lethal mixture for humans. For another, the earth at that time was hot, with average temperatures exceeding 130 degrees F. This was heat from the planet's molten core, produced during the earth's accretion, frictional heating arising from denser materials sinking to the planet's center, and heat from radioactive decay.

During the Archean one of the most important and dramatic events in the history of the planet occurred: the rise of life on earth. Early life took the form of single-cell organisms (and colonies) lacking a distinct, membrane-bound nucleus, the primary examples of which are bacteria, archaebacteria, and photosynthetic forms (like cyanobacteria). That life originated during the Archean means that metabolism, reproduction, and DNA all arose during this period.

The appearance of DNA some 3,900 million years ago makes it the most ancient of all ancient texts. Ancient texts of other types are still revered today, including the 5,000-year-old Yi Ching (2852–2738 BCE), the Bhagavad Gita (Hindu, oral Sanskrit 3137–1924 BCE, written Sanskrit 400 BCE), the Qur'an (Islam, 630 CE, written in Arabic 650–656 CE), Tipitaka (Buddhism, 580–543 BCE, written in Pali in 30 BCE), and the Bible. These texts are widely translated (in up to 2,200 dialects), widely printed (3 billion copies), read, interpreted, downloaded (1.4 megabytes), and even memorized. The Torah has 304,805 Hebrew characters and in the centuries since the original, ascribed to Moses (1444–1280 BCE or Josiah's 620 BCE revision), the number of "mutations" worldwide is only nine (among the Ashkenazi, Sephardi, and Yemenite lineages), all of which are considered to be a result of minor spelling differences that do not impact meaning.

The original ancient text is written in the genomic DNA of every being alive today. That text is as old as life itself, and over 10^{30} copies of it are distributed around the earth, from 5 kilometers deep within the earth's crust to the edge of our atmosphere, and in every drop of the ocean. A version of this text is found in each nucleated cell of our bodies, and it consists of 700 megabytes of information (6 billion DNA base pairs). It contains not only a rich historical archive but also practical recipes for making human beings. For such a significant text, its translation into modern languages began only recently, in the 1970s.

Other naturalistic, geological, and astronomical resources can also be considered ancient texts. We surmise that the ancient texts written by humans, as well as the texts of natural data, are all transmitting profound truths that are not intrinsically contradictory. We try to align and weave these various threads to help us understand the past and the future.

☙ ☙ ☙

Because the engineering of that most ancient text, the genome, takes place at the cellular and subcellular levels, it's important to understand the cell and its workings in some detail. In fact, it would be nice to know exactly what it's like to *be* a cell. But is it possible, even in principle, to know such a thing?

In 1974 the American philosopher Thomas Nagel published a mind-stretching essay that became an instant classic, "What Is It Like to Be a Bat?" The piece was an attempt to understand the subjective character of a conscious experience that is fundamentally different from our own. Nagel found that his ability to do this was rather limited. He tried to imagine having webbed arms, hanging upside down by his feet in an attic, navigating through the air and catching insects by echolocation, and so on. "In so far as I can imagine this (which is not very far)," he said, "it tells me only what it would be like for *me* to behave as a bat behaves. But that is not the question. I want to know what it is like for a *bat* to be a bat."

Certainly we are more like bats (which are, after all, mammals) than we are like cells, and so if it's difficult or impossible to know what it's like for a bat to be a bat, it's going to be an uphill battle to know what it's like for a cell to be a cell. But let's at least give it a try.

First, some history. Cells were supposedly discovered in 1665 by British physicist Robert Hooke, who saw them in thin slices of cork. What he actually saw, however, were not living (or even dead) cells but rather a network of tiny holes arranged in the honeycomb-like structure characteristic of cork, the bark of a dead tree. He thought these small cavities resembled a monk's or prisoner's cell, hence the name. At most, then, Hooke can be credited with coining the term that was later applied to cells in the modern sense.

Genuine living cells were first seen by the Dutch linen draper Anton von Leeuwenhoek through the microscope lenses that he ground as a hobby. (He first used them to examine the quality of cloth samples.) Through his lenses Leeuwenhoek saw one-celled protozoa, blood cells, sperm cells, and many other "animalcules," as he called them. In 1683, pressing against the limit of his ability to discriminate the fine structure of this microscopic underworld, Leeuwenhoek saw bacteria (derived from tooth scrapings), and he vividly described and drew relatively accurate pictures of them.

It wasn't until much later, in the mid-1800s, that the foundational laws of cell theory were stated. In 1837 German botanist Mathias Schleiden made the generalized assertion that all plants are made of similar units called cells. Two years later his physiologist friend Theodore Schwann extended the same claim to animals. Finally, in 1855 the pathologist Rudolf Virchow stated the capstone principle of cell theory, which he expressed in Latin as *omnis cellula e cellula*, which means "all cells arise from cells." (The phrase is reminiscent of Pasteur's *omne vivum ex vivo*—all life from life—and likewise seems to embody a vitalist position.)

In general, cells are small objects. Bacteria range from 0.5 to 750 microns, human cells from 5 microns to 1 meter in length. (A micron is a millionth of a meter; for comparison purposes, a human red blood cell is about 5 microns across.) The longest cells are nerve cells, or neurons, some of which stretch from spine to toe. The very tiniest bacteria, members of the genus *Mycoplasma*, are less than a micron long. The physical volume taken up by a *Mycoplasma* bacterium is evidently the smallest amount of space that will accommodate all of the metabolic machinery necessary for life, or life as we know it, so far, here on earth.

There's one big division in the overall cellular universe: some cells exist only as entities embedded within other cells of the same type, for example, muscle cells or brain cells; other cells exist as free-floating entities by themselves, for example, *E. coli* cells or yeast cells.

What we might call an average or generic cell is composed of eight major classes of polymers: polynucleotides (like DNA and RNA), polypeptides (like collagen and vancomycin), polyketides (like fats and tetracycline), polysaccharides (like cellulose and starch), polyterpenes (like cholesterol and rubber), polyaminoacids (like lignin and polyalkaloids), polypyrroles (like heme and vitamin B12), and polyesters (like PHA, PHV). First in importance are the nucleic acids, DNA and RNA, which contain the genetic information, the software of life. This software runs the cell in as literal a sense as a computer's operating system runs the computer. It directs the formation of proteins. It contains the cell's own recipe, the complete instruction set necessary and sufficient for making another nearly identical cell. The cell is not controlled by its genome exclusively, but also by its environment, its history, and the choices that the cell makes in response to these. Emergent behaviors arise as a function of the cell's being greater than the sum of its parts individually.

Second, cells are made of proteins, which constitute some 20 percent of a given cell by weight. The term "protein" comes from a Greek word that means "primary" or "first thing," and a typical bacterium may possess several thousand different types of them. The proteins perform most of the cell's housekeeping, self-repair, and other workaday tasks. Some of them, the enzymes, or biological catalysts, are shaped with distinctive clefts or pockets that assist in certain chemical reactions. Structural proteins have ends that attach themselves to surfaces to provide rigidity and support to compartmented cells. Transmembrane proteins allow selected types of molecules to enter or leave the cell. Since we are made up of cells, proteins are another thing that cells and humans have in common.

A cell's membrane, which constitutes its outer surface, is composed of lipid molecules. Lipid molecules possess special properties that suit them to the cell's watery interior and to its fluid external environment. A lipid molecule is a long linear structure whose two ends behave differently when immersed in water, which in the case of living cells is most of the time. One end of the molecule, the "head," carries an electrical charge, and this makes that end hydrophilic (water-loving), meaning that it seeks out or orients itself toward water. The molecule's other end, the "tail," is uncharged, which makes it hydrophobic (water-fearing), meaning that it hides or sequesters itself from water. This bipolar feature makes lipid molecules into active, dynamic entities, for when placed in water they will spontaneously associate and collectively adopt the shape suited to the opposed chemical affinities of their ends: generally they will form a sphere, with the outer surface consisting of the charged heads, the inner surface of uncharged tails. Cell membranes are composed of a lipid bilayer, two sheets of lipid molecules sandwiched together.

Collectively, these three cellular elements—nucleic acids, proteins, and the lipid bilayer membrane—exist for the purpose of maintaining the cell

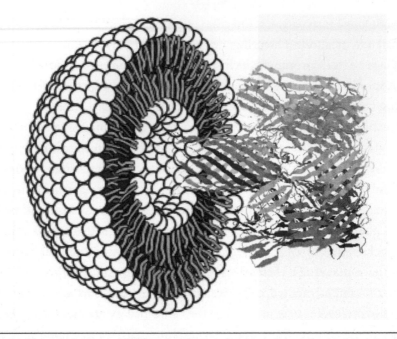

Figure 2.1 Lipid bilayer spherical shell (cross-section) with a protein pore (hemolysin) on right showing how molecules can cross the otherwise impermeable bilayer.

as a living system, but for this it needs a fourth class of materials, sugars (saccharides, typically glucose or sucrose). Sugars are the material sources of a cell's energy. But in order to be actually utilized as energy, they must first be converted into a biochemically accessible form. This conversion occurs in a process called glycolysis ("sugar breaking"), a complex, ten-step progression of events that is the bane of biochemistry students (although other metabolic pathways take even more twists and turns). The final result of glycolysis is the creation of ATP (adenosine triphosphate), the molecules that directly power the cell. Those same ATP molecules power our bodily cells.

ৡ৯   ৡ৯   ৡ৯

To be a cell, then, is to be a deterministic system governed by DNA, composed largely of proteins and lipids, and energized by ATP. Some cells are more like us than you may imagine—*E. coli*, for example, the standard organism of genetic engineering.

This bacterium was discovered in 1885 by the German pediatrician Theodor Escherich, after whom it was named. It has since become the most biochemically well-defined organism known to science. Like ourselves, it is mobile and self-propelled, although the medium through which it travels and its propulsion system are quite different from our own. Individual *E. coli* cells are small, rod-shaped objects about four micrometers in length, easily visible in a light microscope. Extending from the cell's surface are a number of long, corkscrew-shaped flagella. They propel the cell through a watery medium that, to them, is as viscous as molasses. At this scale, where gravity has little effect, there is no up or down. Since the average *E. coli* cell lives inside the human intestinal tract, the cell has no vision, and since it has no brain or nervous system, it has no conscious experience. But believe it or not, the bacterium has a primitive sense of perception.

An *E. coli* cell can be observed swimming through its aqueous medium in a straight line, propelled by its flagella motors. Inside the cell, floating in the cytoplasm, certain kinds of protein molecules react to the density of nutrients in the surrounding medium. If the nutrient levels remain the same or are increasing, the bacterium will continue straight ahead in its travels. But if the nutrient level declines, these same proteins will react with the flagellar motor in such a way that causes the organism to reverse direction. It will then flail about and wander all over, as if exploring the universe, until the nutrient gradient increases, whereupon it will again swim in the direction of the nutrient. In its simple and rudimentary way, this is how the bacterium senses its surroundings.

There is, finally, one other commonality between ourselves and *E. coli*, and that is the ability to reproduce. An *E. coli* cell's capacity for reproduction is unmatched in speed by few other organisms. The basic processes take place in the cytoplasm, which has no precise analog on the human scale. It's a scene of apparent pandemonium where every possible space is filled, all molecules are in a state of constant thermal agitation, and objects of multiple shapes and sizes are bumping into each other, fusing together, breaking apart, and popping into and out of existence seemingly at random.

Biochemist David Goodsell, in his book *The Machinery of Life*, likens the conditions inside the cytoplasm to an airport terminal crowded with passengers who are pushing, shoving, and scrambling in all directions. But this image doesn't really do justice to the interior of an *E. coli* cell. An airport terminal, after all, is a rigid and stable structure that exists in two dimensions (or if we take into account the stratification of arriving and departing as well as domestic and international passengers, and so on, we could attribute to it a slightly larger fractional dimensionality such as 2.1), whereas the cytoplasm is unquestionably a dense three-dimensional throbbing blob. Inside it, however, everything careens relentlessly toward self-replication, for within the space of about thirty minutes, the cell manages to duplicate with extreme precision each and every one of its component parts: its proteins, lipid molecules, even its own genome. And at the end of the process, the cell pinches itself in two, giving birth to a daughter cell clone, which will reproduce itself in another half hour.

The role of *E. coli* in genetic engineering stems from its ability to reproduce itself with its characteristic high speed and great fidelity. According to a standard account (which is probably correct), genetic engineering in the modern sense was born in 1972, when two biologists met for a late-night snack at a delicatessen near Waikiki beach in Hawaii. Stanford University medical professor Stanley Cohen and biochemist Herbert Boyer, of the University of California–San Francisco, were in Honolulu to attend a conference on plasmids, the circular strands of DNA found in the cytoplasm of bacteria. Plasmids can be replicated independently of the cell's chromosomes.

At the conference, Cohen announced that he could insert plasmid DNA into *E. coli* and have the bacterium propagate and clone the plasmid. Boyer described his work with EcoRI, an enzyme (named after and isolated from *E. coli*) that could cut DNA at specific, predefined sites along the length of the molecule. Later that night the two scientists realized that by combining their respective innovations they could splice together fragments of two different plasmids, producing recombinant (mechanically changed) DNA, and then get the bacterium to mass-produce whatever it was that the engineered plasmid coded for.

But as correct as that account may be, Cohen and Boyer were not the world's first (nor even the most successful) genetic engineers. That distinction belongs to viruses, particularly bacteriophages (such as the T4 phage, which looks like a lunar landing module straight from the Apollo program).

A virus is essentially a string of DNA or RNA wrapped in a protein coating. It replicates by inserting its genome into a fully fledged cell, which proceeds to treat this new and foreign set of raw genes as if it were its own original genetic material. Uninfected cells use nucleic acids primarily to make proteins: a molecule of RNA polymerase (an enzyme) unzips a section of double-stranded DNA, reads off its genetic information, and constructs a complementary strand of mRNA (messenger RNA), in a process called transcription. The mRNA is in turn read by ribosomes, along with some other molecular machines, which collectively string together amino acids in the order prescribed by the mRNA, in a process called translation. Since a protein is nothing but a long string of amino acids, when the translation is complete, so is the protein.

Infection by a virus changes the picture entirely. For an *E. coli* bacterium, an attack by a phage virus initiates a cascade of violence and destruction equal to anything offered up by horror fiction. The phages descend on the bacterium's outer membrane like a swarm of bees and forcefully insert their DNA through the cell membrane and into the cytoplasm. The viral DNA invaders now sabotage the cell's normal transcription and translation of the its own genes, and redirect those processes on themselves. At that point, the viral genome has taken over the organism, and some minutes later, with its own enzymes firmly in command of the cell, the virus has been replicated tens to hundreds of times.

The final step—lysis, "bursting"—is the climax of the process. *Under the Microscope*, a collection of photographs taken with electron microscopes, graphically depicts the very moment of the cell's destruction: a malign-looking fleet of new T4 phage particles, each of them an exact duplicate of the lunar landing module virus that started the whole chain reaction, pour out of the bacterium's ruptured membrane in what looks like a microbial-level explosion.

This is genetic engineering as done by nature.

This process is not so different from the way it's done in the lab. As pioneered by Cohen and Boyer, who patented their procedure in 1976, and as practiced by hordes of genetic engineers who followed them, from the ranks of biotech companies to high school students, a gene of interest is inserted into a colony of *E. coli* bacteria, often by means of a plasmid containing the stretch of DNA that codes for the desired substance. To the bacterium, a gene is a gene. So long as the newly inserted DNA segment makes molecular-biological sense (which in many cases it doesn't), the bacterium will produce whatever the gene codes for, and in generally the same way that it manufactures viruses.

Getting *E. coli* to churn out jet fuel is nothing compared to T4 phage viruses getting the same bacterium to fabricate more of themselves; the phages are far more complex entities than simple kerosene molecules.

E. coli bacteria are tiny, obedient molecular factories. You tell them what to make and, by and large, they will make it.

ያ።   ያ።   ያ።

Now we know what it's like to be a cell, particularly an *E. coli* cell, and we understand why it's the vehicle of choice for genetic and genomic engineering. In the Prologue, we saw that with Fred Blattner's Clean Genome *E. coli*, it's possible to improve on and streamline the bacterium's native genetic software. In nature, *E. coli* has 4,377 genes on a genome consisting of 4,639,221 base pairs. Blattner reduced the gene count of the laboratory K-12 strain by some 15 percent, thereby producing an organism that was optimized for laboratory, industrial, and academic research purposes. Within limits, organisms are plastic and malleable, and are susceptible to improvements and efficiencies through intelligent engineering. Indeed, it is a central tenet of synthetic biology that organisms as we find them are not necessarily optimized for accomplishing the various specific tasks to which we might put them.

A series of four experiments performed by Craig Venter at the J. Craig Venter Institute in Rockville, Maryland, together with a lab crew that included the Nobel laureate Hamilton Smith, among others, illustrates the degree to which the genomes of certain organisms can be engineered, modified, whittled down, improved, and even created from scratch, chemically. In 2005 Venter and colleagues attempted to identify the essential genes of a minimal bacterium. Plenty of genes, they found, were not necessary to the organism's functioning, and some even slowed its growth.

The bacterium that Venter and his colleagues worked with was *Mycoplasma genitalium*, an organism that was already known to have the smallest genome, consisting of some 580,000 base pairs, of any known natural microbe that can be grown in pure culture. (The organism is so named because it exists as a pathogen of the human urogenital tract.) *M. genitalium* was also known to have 482 protein-coding genes. The researchers proceeded by selectively disrupting the action of various of these protein-encoding genes, one by one, and observing the effect, if any, on the bacterium. (They did not disrupt any of the forty-three RNA-encoding genes.) By this method they found that one hundred of the protein-encoding genes, or about 20 percent of the total, were nonessential, or as they called them, "dispensable." They also found, possibly surprising to some, that disrupting some genes speeded up the organism's growth rate under certain conditions, meaning that their presence in the organism retarded its growth rate, acting as "some sort of brake on cell growth."

The team members summed up their results: "Under our laboratory conditions, we identified 100 nonessential genes. Logically, the remaining 382 *M. genitalium* protein-encoding genes, 3 phosphate transporter genes, and 43 RNA-coding genes presumably constitute the set of genes essential for growth of this minimal cell." (Still, it is likely that loss of some of these nonessential genes could be lethal in combination. Cases abound where pairs of mutations, each of which is viable separately, lead to death of the organism if both mutations occur in the same genome.) Nonetheless, this experiment showed that the genomes of some bacteria are capable of being drastically changed without damage to the organism, and in some cases even have beneficial effects on their growth rate.

In 2007 Venter's team chemically synthesized the entire 582,970-base pair genome of *M. genitalium*. Since they worked at the J. Craig Venter

Institute, the scientists named their new synthetic genome *Mycoplasma genitalium* JCVI-1.0. The synthesis of this genome was an enormous technical achievement inasmuch as previously assembled synthetic genomes had been much smaller, the next-longest piece of synthetic DNA consisting of only 32,000 base pairs.

In their third experiment, Venter's crew changed one bacterial species into another one. They did this by taking the genome from one species and transferring it into members of the second species, which then turned themselves into members of the first. In this case the researchers used a natural (as opposed to a synthetic) genome, however, and the species in question were two different types of *Mycoplasma*: *M. mycoides* and *M. capricolum.* "These species are more convenient experimental organisms than *M. genitalium* because of their faster growth rate," the researchers wrote in their report on the project, which was published in *Science* in 2009. (*M. genitalium* has an extremely slow growth rate.)

Although the procedure was technically complicated, it was simple enough conceptually, since what the researchers did was to isolate an *M. mycoides* genome and transplant it into wild-type *M. capricolum* recipient cells. For a while there were two different genomes residing in the same cell. Eventually the new DNA was recognized and taken up by the recipient cell, which thereupon transformed itself into an *M. mycoides* bacterium.

"Changing the software completely eliminated the old organism and created a new one," Venter said of the experiment. This might at first glance sound like a magical changeover, but the invading genome was merely acting like a virus, taking over and transforming the cell into which it had been placed. Just like Venter's genome, a virus is software that completely eliminates the old organism and creates new ones.

Still, Venter's capping and culminating experiment was yet to come. This was to design, digitize, and then chemically assemble a 1.08-million base pair *M. mycoides* genome and boot it up inside a cell. They called this synthetic genome *M. mycoides* JCVI-syn1.0. Then they did with it exactly what they had done with the natural *M. mycoides* genome of the earlier experiment: transplant it into an *M. capricolum* recipient cell. The results were the same: the new (synthetic) genome took over the old *M. capricolum* cell and turned it into an *M. mycoides* cell.

As the researchers told the story in *Science*: "There was a complete re-placement of the *M. capricolum* genome by our synthetic genome during the transplant process. . . . The cells with only the synthetic genome are self-replicating and capable of logarithmic growth."

ℬ ℬ ℬ

These developments created a minor sensation in the scientific world and a major sensation in the general media ("Scientists Create Artificial Life"). There were news reports saying that President Barack Obama had ex-pressed unspecified "genuine concerns" about this work.

With one notable exception, however, Venter and his colleagues were quite restrained in their claims. In their report on the project, the re-searchers drew two general conclusions from what they had done. First, "the demonstration that our synthetic genome gives rise to transplants with the characteristics of *M. mycoides* cells implies that the DNA sequence upon which it is based is accurate enough to specify a living cell with the appropriate properties." In other words, there are no mystic features, holdovers, or leftovers from vitalism pertaining to DNA molecules. Whether they were "natural" or "synthetic" genomes, they still controlled a cell.

Second, "this work provides a proof of principle for producing cells based on genome sequences designed in the computer. DNA sequencing of a cellular genome allows storage of the genetic instructions of life as a digital file." The reduction of genetic instructions to a digital file delivered a knockout second blow to vitalism.

But then the scientists advanced a third claim: "We refer to such a cell controlled by a genome assembled from chemically synthesized pieces of DNA as a 'synthetic cell,' even though the cytoplasm of the recipient cell is not synthetic." They made it sound as if they had created an artificial life form even though a nonsynthetic, natural cell had actually given rise to the new organism.

The genome constitutes only about 1 percent of the dry weight of a cell, which means that only a tiny proportion of the cell is actually syn-thetic. The rest of the organism was as natural as any other ordinary cell. Indeed, Venter's synthetic genome depended on the rest of the recipient cell's natural and native apparatus for its expression: it depended on the cell's molecular machinery of transcription, translation, and replication, its ribosomes, metabolic pathways, its energy supplies, and so on. (Although Venter was fond of saying that "the DNA software builds its own hardware," it would be more accurate to say that the recipient cell builds whatever hardware the DNA software codes for—and only if the existing hardware is pretty close to the target hardware already.)

Building a living cell that is genuinely synthetic is one of the goals of synthetic biology. By separating what's essential to living systems from what's not, such a cell would advance our understanding of what constitutes the necessary and sufficient conditions for being alive. In addition, a synthetic cell, provided that it is also a *minimal* cell, is considered by some to be a beguiling platform for genomic engineering since its lack of extraneous or inhibitory components might improve its efficiency at turning out desired end products such as biofuels, medicines, vaccines, or green chemicals, although others say that a larger genome is better.

Further, discovering or creating a minimal organism would establish the limits of what's possible in the miniaturization of living systems. Biological minimalism can exist on two different levels. First, there is the minimal *genome*: the smallest genome that is sufficient to create, maintain, and replicate itself. Possibly such a genome could be as small as two 3-mers (three-part molecules) that come together to form a 6-mer (a 6-part molecule). Increasingly interesting genomes (of 187, 2587 and 113,000 base pairs) will be introduced soon (explained below).

On the second level there is the minimal *cell*, composed of the fewest components that can jointly carry out all the normal processes of life, including metabolism, reproduction, and evolution. As the genome grows larger, it gets steadily harder to separate the full length of the two strands in order to make new copies. The solution is to separate out only a little at a time (say a dozen base pairs out of millions) and synthesize a new strand of DNA or RNA a few base pairs at a time—with a long copy emerging from the intact double helix. This argues for a separation of information storage and factory functionalities. These functions are typically encoded in RNA and in proteins that fold up into complex machines instead of being long rods. The ability to fold gives access to vast capabilities, and in principle this could be done simply with RNA genomes and with RNA as folded machines. But RNA has only four closely related functional groups (A, C, G, U), so the coding of another class of polymers (proteins) with vast diversity (20+ amino acids) was, perhaps, inevitable.

Back to what it's like for a cell to be a cell. Animations help us visualize how polymers are made from monomers. Often these are depicted as happening in an orderly fashion, similarly to how workers on an assembly line might pass car parts down the line "just in time" for the next sequential production step. The process, however, is hardly so orderly. In reality, the four nucleotides or twenty amino acids are randomly tried out before a single correct one is accepted, and each of these with considerably more jostling and false moves than is typically shown in animations.

Seemingly, a minimal genome would automatically produce a minimal organism, but this is by no means obvious. Some protozoa, for example, have genomes that are over one hundred times larger than the human genome, which means there is a big mismatch between the size of the genome and the size of the corresponding organism. This is largely due to the fact that large stretches of the protozoa's genome may consist of noncoding regions (sometimes called junk DNA), meaning dispensable under some circumstances. A minimal genome, however, would exclude such sequences by design and intention. Still, a cell produced by such a genome might nevertheless contain extraneous, redundant, or other inessential components. Whether a minimal genome will in fact produce a minimal cell is something that can be decided only after the fact, by experiment, not in advance, by theory. (But if "theory" means, as it often does, going from vast numbers of experiments to a new best guess, then the minimal genome will likely come from theory.)

Attempts to build a synthetic cell have not been entirely successful. In 1969, for example, three biologists at the State University of New York–Buffalo, K. W. Jeon, I. J. Lorch, and J. F. Danielli, decided to create a synthetic living organism. "After participating in a symposium on the experimental synthesis of living cells," they wrote in their report on the project in *Science*, "we decided that we had the means to carry out the reassembly of *Amoeba proteus* from its major components: namely, nucleus, cytoplasm, and cell membrane."

Amoeba proteus is a comparatively large (0.4 mm) aquatic organism that is easy to work with using tools such as micropipettes and other micromanipulators. And so the experimenters took the nucleus from one amoeba, the cytoplasm from a second, and put them together inside the evacuated cell membrane of a third. Eighty percent of the time, the new composite organism lived. "The techniques of cell reassembly appear to be sufficiently adequate so that any desired combination of cytoplasm, nucleus, and membrane can be assembled into living cells," the researchers concluded.

Unfortunately this cobbled-together organism was not really a synthetic cell, for all the parts used were natural; only their locations had changed. It was more like reshuffling the deck than providing the players with a new set of cards.

A truly synthetic cell is one that we create ourselves, from the ground up. This could be a new form of living matter fabricated out of pure ingredients. Such a cell might tell us something about the original cells that arose at the dawn of life on earth. Arguably, life originated when a group of molecules and molecular structures first organized themselves into living systems, but precisely what those molecules were and how they arranged themselves so that life emerged from the mix is an open, possibly unanswerable question. Nevertheless, successfully creating a synthetic cell would represent a key advance in the understanding of living processes. For life, like a machine, cannot be understood simply by studying it and its parts; life, to be understood, must also be put together from its parts.

Яⷧ   Яⷧ   Яⷧ

How then do we create a truly minimal living cell that is also genuinely synthetic? In 2006 a colleague and I advanced a proposal for the creation of a minimal cell that in addition would be a substantially synthetic living organism. My colleague is Anthony Forster, then of Vanderbilt University

Medical Center, now of the University of Uppsala, and our object was to build, from the molecules up, a chemical system capable of replication and evolution. As opposed to Venter's reductionist biology, this would be an example of *constructive* biology, the putting together of a living entity from its constituent parts.

The work would proceed by starting with the smallest molecular components and arranging them into subsystems, then having those subsystems self-assemble into larger units, and so on.

We designed our genome by looking for genes that have homologues, closely related sequences, in the genomes of several groups of organisms. The rationale is that genes that appear in many groups of organisms are somehow "required" for those species. This method, along with others from genetics and biochemistry, suggested a genome that consists of only 151 genes and is only 113,000 base pairs long. Our plan is to construct the genome, place it inside a lipid-bilayer membrane-sphere filled with the macromolecular enzymes encoded by the 151 genes and a minimal inventory of small molecules needed for life. The entire system could finally be bootstrapped into existence by the addition of synthetic ribosomes, translation factors, and other structures inspired by similar components existing in natural cells such as *E. coli*.

Eventually this approach will produce a synthetic, self-replicating, and self-sustaining minimal cell. To prevent the cell from replicating outside the laboratory, we are building into it a deliberate dependence on nutrients not found outside the lab environment.

Unlike the synthetic bacterial genome, where many genes are of unknown structure or function, the end result would be a functionally and structurally well-understood self-replicating biosystem, synthetic but nevertheless alive and well. If and when it happens, it will be a major milestone in the history of biology: civilizing, taming, and domesticating the basic processes of life. This synthetic minigenome presents a clear path to the mirror world introduced in Chapter 1. The synthetic minimal cell would enable the production of materials too large or otherwise incompatible with the more elaborate functioning systems of a complex cell. It also represents our best shot at a general nanotech assembler, the dream of Eric Drexler and many nanotechnology enthusiasts since he first described it in his 1986 book *Engines of Creation*. We could then harness these synthetic minimal cells and put them to use in drug, vaccine, chemical, and biofuel development.

CHAPTER 3

-500 Myr, Cambrian

The Mirror World and the Explosion of Diversity.
How Fast Can Evolution Go and How Diverse Can It Be?

☙

Life, the Genetic Code, the Mirror World, and the Generation of Diversity

Mirror life, including mirror humans, may sound like something out of science fiction, an outlandish concept unrealizable in practice even if conceivable in theory.

But mirror life is a real possibility, not just a flight of fancy. To convince you of this I'd like to show you how it can be created. But first we need a deeper understanding of life itself, in all of its complexity.

What, then, is life?

Erwin Schrödinger's short 1944 book *What Is Life?* inspired physicists such as Maurice Wilkins and others to establish the field of molecular biology. Schrödinger championed the idea of life as based on an "aperiodic solid," a suggestion that anticipated DNA as the sequenced biopolymer in which the genetic information is encoded.

Many people think that life is an all-or-none, black-or-white, on-or-off, matter-antimatter phenomenon, with no in between. However, let's consider the possibility that life is a continuous, scalable, and measurable property. Similarly, many thinkers are tempted to argue that life is "the pinnacle of complexity." But let's consider replacing complexity with the notion of replicated complexity (which can be shortened to "replexity"), or mutual information. Two images composed of scads of random ink dots seem equally complex, equally likely or unlikely. Similarly, the atomic arrangement of two stones may seem equally complex. But if we see an image of a complex set of dots with a mirror duplication (like a Rorschach inkblot) or a "living stone" (*Lithopsjulii*), we experience the feeling that so much information is unlikely to be duplicated or transformed in a predictable manner from page to page, leaf to leaf, or cell to cell within those leaves. Two complex random patterns seem unremarkable consequences of an inorganic, lifeless world, but two complex patterns that look precisely alike are a hallmark of life.

From an information theory standpoint, we need fewer bits by far to convey an image of a grain of salt and an ice crystal than a movie of a liquid mixture of the two. The chaotically changing solution is more complex than either of the two crystals. Complexity increases with chemical randomness or entropy. Chemical systems tend to spontaneously go toward random, complex mixtures. In information science, entropy refers to the number of bits needed to transmit a complex message. Both from a chemical and an information entropy viewpoint, a living system is less complex than the solution but more complex than the crystals. It would have higher replexity than both because the living system requires more bits to describe its repeating motifs (e.g., its nearly identical DNA, proteins, and cells) than the strict and simple repeating motifs of the crystals or the disordered and weakly repeating structure of the solution. So too the chemical imperative for randomness is satisfied in the solution.

The notion of replicated complexity directly addresses seven counterexamples that have challenged previous definitions of life.

(1) Life as a growing and replicating system. Fire, then, may seem to be a living thing, as it can grow and replicate itself. It can even "evolve" to have new properties; for example, while devouring a succession of materials with different flash points (gasoline at -40 C, ethanol at 13 C, diesel at 62

C, vegetable oil at 327 C, Mg metal at 634 C). Indeed, a flame can replicate in a wide variety of environments while many endangered species can't survive even with tender loving care in a zoo. However, a flame has little long-range order and so would have a low replexity value.

(2) Replexity also removes difficulties with apparently living things such as mules (the sterile offspring of a male donkey and a female horse) that normally cannot reproduce and thus lack a defining characteristic of life: the ability to self-replicate. But mules have reasonably high replexity for three reasons. First, they have a non-zero probability of whole-organism replication. Second, their cells and subcellular components replicate. And third, they are closely related to organisms that do replicate (i.e., their parents).

(3) Life as a steadily increasing reservoir of replexity. A lone herbivore could destroy a complex plant ecosystem and then die, resulting in a net decrease in replexity. This is an average property, not a guarantee. This is analogous to the laws of thermodynamics, which describe the average behavior of vast numbers of molecules, not the instantaneous motion of a few molecules.

(4) Should life display motion? "What about a frozen animal?" you might ask. "That doesn't seem very alive." Many complex organisms survive a freeze-thaw protocol (e.g., human IVF embryos and *Chironomous* larvae). In principle, a frozen organism could be assembled one molecule at a time with small tweezers such as those of an atomic force microscope. (This has not been done yet, but I addressed the idea of assembly from parts among the five grand challenges to vitalism discussed in Chapter 1 and in the context of booting up minimal genomes in Chapter 2.) If we wanted to email a description to specify that organism reliably, we could determine the minimum number of bits needed to transmit the specifications of the organism's structure so that it would be in a living state on construction and warming. The replexity of frozen animals reflects both their origin from a replicating entity as well as probability of replicating again on thawing.

(5) Life as machines (e.g., photocopiers or 3-D printers) that make copies of unique complex objects. So far, these cannot self-replicate. We

can make "trivial" self-assembling robots that simply catalyze assembly
from two (or a few) already complex, nearly functional partial robots. But
here the increase in replexity due to replication is minuscule and the inherent replexity of the parts is due to life (indeed, the intelligent human
life that created the partial robots). There is no real barrier to someday
having 3-D printers that replicate and evolve. Indeed these properties may
be useful for space missions, and due to the vast yet harsh resources, such
printers could become considered more alive than all of earth-bound life
combined.

(6) Life as translation from one polymer type to another (e.g., from
RNA to proteins or from computer bits to DNA). How do we weigh the
significance of such translation? Various precise translations are more impressive than replicated complexity. We can measure the amount of structural identity between two polymers—something called homology by
computationally inclined biologists. In most living beings, the oligonucleotide AUGUUU would encode the dipeptide Met-Phe. A mutated copy
AUGUUG would be called homologous to AUGUUU, while the relation
between AUGUUU and Met-Phe would constitute mutual information
and not homology, since those two molecules differ structurally; indeed
they are related only by a translational code and a molecular machine that
embodies that code. A large amount of mutual information relative to homologous copies found in comparing two objects constitutes evidence for
increasingly advanced and adaptable life. In other words, statistically significant connections (indicative of causal connections) among diverse and
detailed objects is one of the most impressive aspects of life (and of intelligence as well).

(7) What about the repeating motifs of geological strata or inorganic
mold replication processes? We might find hundreds of mineralogically
distinctive layers, some globally replicated (e.g., the iridium layers indicative of a large global dust cloud from the largest meteors to hit earth). But
the precision of patterns and placement is weak (a roughly millimeter to
meter scale). The fossils found in those layers can have micron-scale precision both for the surfaces of the solid "negative" molds of originally complex plants and animals and then the "positive" replica formed by inorganic

minerals replacing the decaying organic cell structures. Nearly all of the replexity in this case is due to the original replexity of the trapped living beings. And even given that, the replexity has diminished by at least a billion-fold from cubic nanometer– to cubic micron–scale replicas in the imprecise mineralization process. (Positive/negative molds and the profound idea of structural complementarity are discussed in Chapter 1.)

As we will see in Chapter 7, there are six industrial revolutions that were intimately connected with the development of relevant quantitative measures. Replexity may be part of the current revolution and the sense of awe we feel as we learn more about what life is.

৯৯   ৯৯   ৯৯

Genome engineering will allow us to become ever more diverse, enhancing our prospects for survival. We are already extending the bodily properties of our species in several different ways. For example, we are making attempts to improve our health, lengthen our life span, and boost our immune system and disease resistance, among other things. We can also adapt to very high and very low population densities as occurs in space exploration (see Chapter 5 and Epilogue).

The replexity (replicated complexity) involved in efforts to create a mirror world and increase diversity is vast, and exists not in the form of arbitrarily diverse or complex artifacts, like a Jackson Pollock splatter painting or a Rubik's Cube, but rather in the form of thousands of identical (or nearly identical) biomolecular devices that are themselves the survivors of the many rounds of production, change, testing, and rejection inherent in the processes of evolution—reproduction, mutation, selection, and extinction. One of the primary errors of eugenics was the search for an optimal genetic specimen. Even if there were such a thing, it would be at risk, because one of the main lessons that we learn from nature is that species that are lowest in diversity are most vulnerable to extinction.

Consequently I am proposing a vision of maximal genomic diversity. But how is that vision connected with the minimal genome of Chapter 2? I will first show how the minigenome enables a huge expansion of replexity, and then address how we might adapt and even outdo the tricks that cellular, then multicellular, and finally multi-organismal (social) systems exploit in their never-ending quest for diversity.

The ancient minimal genome, whatever its exact composition, gave rise to the totality of life forms that we see before us today. Still, there is the problem of how that genome was able to produce the complex set of amino acids, peptides, proteins, and finally the full range of living organisms that now exists. This raises the chicken-and-egg question of how life got started in the first place.

The problem is that modern cells are based on both DNA and proteins. But in order to replicate, DNA needed the proteins which act as enzymes that catalyze the whole replication process. On the other hand, the proteins themselves could not exist without DNA molecules, because those very DNA molecules contain the recipes for making the proteins. So, which came first: DNA or the proteins?

The RNA world probably came first. RNA is essentially single-stranded DNA (with uracil in place of DNA's thymine and an extra oxygen on each sugar). Because RNA molecules are able to act as enzymes, the prevailing theory today is that if "in the beginning there was RNA," then RNA molecules could have catalyzed the basic reactions needed for life. Such catalytic RNA molecules are known as ribozymes (enzymes made of ribonucleic acid).

To review what we learned in high school, RNA molecules come in a variety of types, which collectively constitute the basic toolbox mechanisms of protein synthesis and cellular replication. Two types of RNA are especially important. The first is mRNA (messenger RNA), which is a single-strand RNA chain that carries the genetic information from the DNA in the cell's nucleus to the ribosomes in the cytoplasm. That information is carried in the form of a (long or short) sequence of nucleotide bases. Another crucial class of RNA is tRNA (transfer RNA), small (~75 nucleotides long), folded molecules that transport amino acids from the cytoplasm to the ribosome. These two types of RNA meet up in the ribosome, where the tRNA amino acids get strung together in the sequence dictated by the mRNA, to form a protein. (One component of a tRNA molecule consists of a triplet of nucleotide bases that code for, or specify, a distinct amino acid [as, for example, UUU codes for the amino acid phenylalanine]. Thus you can regard tRNAs and amino acids as matched sets: each tRNA goes together with the specific amino acid encoded by its three-letter codon.)

The genetic code bridging the RNA world and the protein world was probably in place and fairly optimal before the core code protein enzymes existed. There were possibly fewer than twenty tRNA molecules and catalytic rRNA (ribosomal RNA), possibly shorter than today's rRNA molecules, plus an RNA replicase (an enzyme that catalyzes the synthesis of a complementary RNA molecule using an RNA template). The replicase might be as small as the recent 187-mer ribozyme capable of copying RNA molecules as large as 95-mers (but it cannot yet copy itself).

The smallest peptidyltransferase ribozyme (which establishes links between adjacent amino acids of a peptide) can be embedded in the tRNA structure itself. The smallest ribozyme capable of covalently attaching amino acids onto tRNAs is 45-mer—a ribonucleic enzyme that's forty-five units long. This means that a minimal ribo-world (or RNA world) just ready to explode into the ribo-peptide world (or RNA-world-plus-peptides) might have been a bit bigger than $187 + 20x (75 + 45) = 2{,}587$ nucleotides long (where 75 is the average tRNA length and 20 is the number of tRNA molecules). The next step along the path to life might have been to a polymer similar to the 113 kilo base pair-long minigenome described in Chapter 2.

How can we measure this evolutionary "progress"? Some say that the idea of progress is delusional and that the phrase "survival of the fittest" is tautological. But how about measuring evolutionary progress by its degree of replexity (which, remember, stands for replicated complexity)? The transition from the RNA world to life extends informational complexity to bioinformational replexity. One nearly perfect copy of any given complex structure suffices to qualify it as a candidate for replexity. Making additional copies doesn't increase replexity further, except to the extent that the copies are imperfect and hence capable of small and large impacts on selection. In Chapter 2, we saw that synthetically making a copy of a genome is not as promising for changing the world, or our view of it, as is making a radically new genome—or better yet a set of such genomes. Photographing the Mona Lisa is not as impressive as creating it in the first place.

Above I spoke of a truly minimal 187-mer replicase, and then of a slightly larger genome (2,587 bases) capable of translating from the RNA world to an RNA-plus-protein world. The next step up the ladder of replexity is to an even larger genome, one capable of high-fidelity replication and using DNA in place of RNA. In going from 187 to 2,587, and then to 113,000 bases, we see more than just an increase in complexity of the genome. We see an expansion of the alphabet from four bases to twenty more letters in the form of amino acids. (Each amino acid has its own one-letter code; phenylalanine, for example, is F, leucine is L, and so on.) This addition of a new polymer class (i.e., amino acids strung together into proteins) greatly expands the range of folding, catalysis, and functionality in general. The two standard base pairs and occasional nonstandard pairs (in Figure 1.4) of the RNA world basically look and act chemically neutral—flat inert plates. The twenty amino acids embody new positive, negative, and variable charges as well as highly reactive sulfur atoms that are all foreign to the A, C, G, U nucleotides of RNA. Furthermore, proteins have roughly $20*20 = 400$ possible amino acid pairings ($20*20/2 + 10 = 210$ unique pairings, actually—including self-pairs), each pair having a greater diversity of interaction angles and chemistries than RNA possesses.

In line with our single-minded quest to increase diversity, it's relevant to ask how we could expand this raw treasure trove of replexity in the future. Just adding mirror nucleotides and mirror amino acids would essentially double the number of these basic units (monomers) from twenty-four (4 nucleotides + 20 amino acids) to forty-seven (by adding 4 more ribo-nucleotides and 19 more amino acids [not 20, however, since, as noted in Chapter 1, glycine is so simple that it is its own mirror image]). These mirror monomers combine into linear polymers (RNA and proteins), and even though they have radically new characteristics, these RNA and protein polymers are nevertheless completely predicable in terms of their binding and catalysis properties so long as we know the properties of the original (nonmirror) versions.

Then, as an even more radical advance in complexity than creating a separate world of mirror amino acids, we could create proteins made of a mixture of *both* types of amino acids, the standard twenty plus the nineteen mirror versions, for a grand total of thirty-nine. (As we saw just above, since glycine is its own mirror version it does not constitute an additional amino acid.) For proteins of length N, we could make 20^N possible chains using the standard set and 39^N possible proteins using the expanded, "mirror" set of amino acids. Initially we would keep the systems either pure standard or pure mirror, since putting both types of amino acids into a cell without the benefit of more knowledge than we have now would almost certainly be toxic to a standard living cell. Why? Well, we know that as a result of billions of years of evolution and selection, pure standard proteins work correctly, and we know that pure mirror proteins fold predictably, but a mixture of the two would yield truly new folds, which we can't yet reliably predict. So for the time being we have been limited to *either* the twenty standard amino acids *or* the nineteen additional mirror versions, and could not use a mixture of both.

On the top left of Figure 3.1, the amino acid F is attached to the only appropriate tRNA using an amazing minimal 47-mer synthetic ribozyme capable of catalyzing the addition of any amino acid to any tRNA. In a cell, twenty protein enzymes (synthetases) discriminate among the thirty-two tRNAs to bond the correct amino acid to the appropriate tRNA. But in the lab, by contrast, the ribozyme is specific only if a researcher is putting the correct pure amino acid and pure tRNA in the same reaction tube.

And here we come to the climax of the mirror-world creation story. For this very promiscuous nature of the artificial ribozyme bonding amino acids to RNAs—the same ribozyme catalyzes the bonding of *any* amino acid to *any* tRNA—represents the key step in the plan to construct the mirror world introduced in Chapters 1–2. The artificial ribozyme will accept any amino acid, including even mirror amino acids, while the natural synthetases have evolved to avoid the use of more than one amino acid— including mirror amino acids.

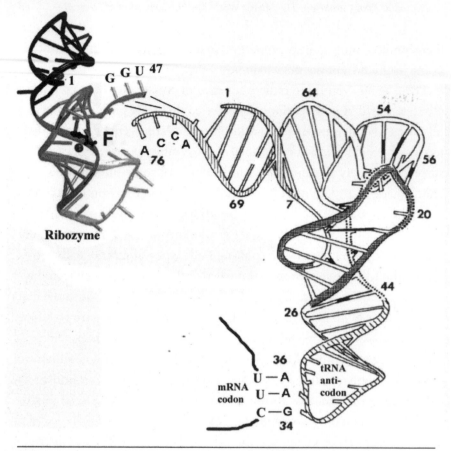

Figure 3.1 The genetic code is being read and decoded at every moment in every living cell on earth. There are two key steps in this decoding process. The first step is the bonding of each amino acid to an appropriate tRNA, and the second step is having that tRNA lining up on the messenger RNA (mRNA) conveyer belt in the ribosome, one triplet codon at a time. The first step is illustrated above with the amino acid phenylalanine (F) being added to the end of the tRNA (at position 76) under the catalytic direction of a synthetic (unnatural) ribozyme (upper left). The ribozyme mainly recognizes the tRNA by three base pairs: ACC (73 to 75) of the tRNA, and GGU (45 to 47) of the ribozyme, by the base pairing rules (A:U and C:G, in Figure 1.4). The second step is visualized in the lower right where three base pairs form between an mRNA codon (UUC) and a tRNA anticodon (GAA 34 to 36), again by the same base pairing process (Figure 1.4). (An anticodon is the sequence of three nucleotides in a tRNA molecule that corresponds to a complementary codon in an mRNA molecule.) The backbones of these two interacting RNAs are long curved lines, and the base pairs are represented as short straight lines.

As I mentioned in Chapter 2, each amino acid typically bonds to one, two, or three of the thirty-two tRNA types, depending on the amino acid (as preparation for orderly assembly of them into long protein polymers). Any deviation is a serious error often resulting in the whole protein being junked by the cell (300 amino acids lost because of just a single wrongly positioned one!).

The mirror amino-acid-bonded tRNAs then constitute the "food" that a standard ribosome with standard mRNA can employ to make mirror proteins. Each tRNA is matched with mRNA, three base pairs at a time. The mRNA ratchets along through the ribosome (in units of these base triplets) directing the addition of further mirror amino acids to the growing protein. And at the end of the process, *voilà!* A mirror protein!

The anticodon from position 34 to 36, in this case GAA, binds with UUC. It can also bind to UUU, but binding to any other triplet would be a mistake and the ribosomal machinery has evolved to avoid this at all costs. The error rate of a ribosome reading triplets and adding the specified amino acid is less than 1 per 100,000. As it happens, the very first codon ever decoded was UUU, which coded for phenylalanine (F). In 1961 Marshall Nirenberg and Heinrich Matthaei noted that long stretches of UUU (poly-U) encode long polymers of F (poly-F). Moreover, the very first folded RNA gene product was the tRNA encoding F in 1977 (from the Rich, Kim, and Klug laboratories, assisted by a teenage version of myself). So, the molecules shown in Figure 3.1 are historic—and also futuristic since the remarkable synthetic ribozyme pictured allows us to tear down the wall limiting us to the standard twenty or so amino acids in proteins ever since the code was formed at the dawn of life. All due to multiple forces maintaining the status quo, including metabolic enzymes, synthetases, and ribosomes that have evolved to destroy or avoid nonstandard amino acids.

To accomplish the transitional magic above (from standard to mirror amino acids), and to move beyond them to a fully mirror biological world, requires more work and more parts. Lots more work and parts. It requires more than just twenty tRNAs to accurately translate forty codons; you need thirty-two tRNAs to translate sixty-one codons, plus two release factor proteins to translate three stop codons; plus the ribosome consisting of fifty-four proteins and three RNAs, plus the initiation and elongation factors to start the translation and keep the mRNA ratcheting along. Oh, and we also need DNA and RNA polymerases, plus a few enzymes that modify the RNA, and proteins to make them more efficient and accurate. This project was briefly introduced in Chapter 2. All of these parts add up to 113,000 base pairs, total, in the smallest minigenome capable of protein-based life, as Tony Forster and I laid out in a *Nature MSB* paper in 2006. This is *five times smaller* than the smallest free-living genome (*Mycoplasma*). Every DNA, RNA, and protein molecule on this minigenome parts list would need to be chemically synthesized in a mirror version (all of the mirror monomers are already commercially available) or to be made by some cute transition trick like the one in Figure 3.1.

⚘   ⚘   ⚘

Just as life leads us to the enhancement of diversity, we now move from the question of how we achieve the minimal replexity capable of new functions, to the bigger question—What are the maximum limits to (engineered) replexity? The above very compact minigenome will get us through the bottleneck of going from the RNA world to the RNA-plus-proteins world. But for most other purposes we want robust, complex metabolic and regulatory systems. This means that we need much more than the proteins required to make proteins from amino acids: we also need the protein enzymes to catalyze the synthesis of amino acid monomers from minerals—and we also need thin air! The enzymes mentioned transform the atoms in sulfate and phosphate minerals, whereas the nitrogen and carbon dioxide of thin air are transformed into the atoms of amino acids and nucleotides. And as if that weren't enough, yet more proteins are needed to scavenge and horde resources, to wage war on other beings, and so on. Typically each does only one job well, so we need thousands of proteins per genome, and billions of proteins per ecosystem.

Ideally, for doing metabolic engineering—making exotic biomaterials, pharmaceuticals, and chemicals—we would have all possible enzymes encoded in one master genome and we'd be able to turn their expression on and off, up and down, like a light switch, as needed for various tasks. So instead of a minimal genome, we want a *maximal* genome—like the vast online shopping world as opposed to the bare survival backpack. We call this collection of nearly all enzymes *E. pluri*, as in *E. coli* meets *E pluribus unum*. "From many states to one nation" becomes, in the world of genome

parts of the protein or RNA chemistry while the side chains (groups of molecules attached to the backbones) provide the variation (the spice of life). Normal (meaning naturally occurring) backbones consist of alternating sugar and phosphate molecules bonded together in an extended chain. Can we utilize totally new backbones? You bet! Another key polymer backbone is the polyketides—polymers of monomeric units consisting of two carbons plus one oxygen (and various derivatives of this) found in lipids and antibiotics.

Like RNA and proteins, the polyketide polymer is colinear with the basic information in DNA, meaning that the order of base pairs in the DNA determines the order of monomer type in the other polymers. The chemotherapeutic drug Doxorubicin is biosynthesized as a 10-mer (10 ketones in a linear polymer) with the sixth position modified from a ketone to an alcohol. In going from DNA to protein, the number of letters drops by threefold (e.g., UUU becomes F [phenylalanine]). In principle, we could make a ribosome that translates from mRNA directly to polyketides (instead of directly to polypeptides), but in nature this is accomplished in a more indirect manner by going from polynucleotide (mRNA) to polypeptide (enzyme) to polyketide. Each ketide monomer addition step requires its own enzyme domain (and these domains are each about 500 amino acids long), so the number of letters gets compressed down another 500-fold. This may seem verbose, but it gets the job done. Still, the possibility of constructing a more general and less verbose version on the ribosome is being explored in my lab.

Polyketides turn into long carbon chains (like fatty acids) and flat aromatic plates (like the tetracycline family of antibiotics). Long carbon chains and aromatics are the key components of diamonds and graphene, which have some truly amazing material properties—for example, single-electron transistors, and/or the strongest and stiffest, have the highest thermal conductivity of any materials yet discovered (as recognized by the 2010 Nobel Prize for Chemistry). This class of biopolymer has many uses already when harvested from cells, and has even more profound applications in the future in biological and nonbiological systems as we get better at designing, selecting, and manufacturing this class of molecules.

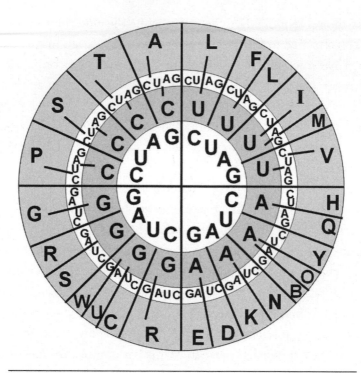

Figure 3.2 The translational code. Four RNA bases (U,C,A,G) make sixty-four triplets (3 base sequence in order from the innermost ring). 4*4*4 = 64. Each triplet encodes one of the twenty amino acids (they appear in the outermost ring: A,C,D,E,F,G,H,I,K,L,M,N,P,Q, R,S,T,V,W,Y). In addition, a few of the triplets encode three stop codons (UAA, UAG, UGA), which tell the ribosome to stop synthesizing whatever protein molecule is currently being made. These stop codons are often called by their historic names ochre, amber, and umber, and are abbreviated here B, O, and U, respectively. (O and U codons encode the less common amino acids pyrrolysine and selenocysteine, respectively, in some organisms.) In the early 1960s, members of the Steinberg, Epstein, and Benzer labs at Caltech were studying viruses that infect some strains of *E. coli* and not others. These researchers offered to name any mutant that might have these amazing properties after the one who found it. It just so happened that a student named Harris Bernstein (who normally worked on fungi, not viruses) found it. "Bernstein" in German means "amber," and so they named the UAG stop codon amber. They later dubbed the other two stop codons ochre and umber to maintain the color theme. To let everyone know how hip you are, you can purchase (at thednastore.com) a bumper sticker that says "I Stop for UAA."

So what have we learned in this discussion of the mirror world, the genetic code, and the generation of diversity? Basically, the following lessons: (1) It is indeed possible to create mirror amino acids and mirror proteins with predictable (small- and large-scale) properties. (2) With the addition of more work and more parts, we can create a fully mirror biological world. (3) These, along with new amino acids and backbones, can then give us access to an entire new world of exotic biomaterials, pharmaceuticals, chemicals—and who knows what else.

How Fast Can Evolution Go?

As we continue our crusade for increased diversity, we must now put the essential question: How fast and how diverse can evolution be made to go? We can increase diversity and replexity by adding genes and by adding needed polymer types, including but not limited to mirror versions of standard chemical reaction types and totally new reaction types, but how does evolution scale up to make the truly marvelous functional diversity in the world? The answer lies in the key parameters involved in maximizing the rate of evolutionary change: population and subpopulation sizes, mutation rate, time, selection and replication rates, recombination (the rearrangement of genetic material), multicellularity, and macro-evolution (evolutionary change that occurs at or above the level of the species).

Worldwide we have a steady state of 10^{27} organisms and a mutation rate of 10^7 per base pair per cell division. The Cambrian explosion happened in the brief period from 580 to 510 million years ago, when the rate of evolution is said to have accelerated by a factor of ten as seen in the number of species coming and going, and as defined by fossil morphologies. The rearrangement of genetic material (recombination) occurs at a rate of about once or a few times per chromosome in a variety of organisms. On a lab scale we typically are limited to 10^{10} organisms, and less than a year for selection, so we are at a 10^{17} and 7×10^8-fold disadvantage, respectively, compared to what occurs in nature. So if we want to evolve in the lab as fast as nature did it during the Cambrian, or even faster, then we have to make up for these disadvantages with higher mutation, selection, and recombination rates.

The fastest lab replication times for a free living cell are held by *E. coli* and *Vibrio,* which clock in at about eleven minutes and eighteen minutes per replication, respectively, speeds that are probably limited by the rate at which a ribosome can make another ribosome. In principle, even though a bacterial virus or phage might take sixteen minutes to complete an infection cycle, the burst size (in terms of the number of virus [phage] babies released when the host cell bursts) is 128, and so the doubling time of the virus is $16/7 = 2.3$ minutes (since $2^7 = 128$).

Other replication speed demons are flies (midges and fruit flies) due to their boom and bust lifestyle. Flies crawl around starving. Then suddenly a piece of fruit appears, and the flies that manage to convert that piece of fruit into fly eggs fastest win. The best flies lay an egg every forty minutes. But even more impressive, some of its genomes—indeed whole nuclei (which look a bit like cells surrounded by a nuclear membrane)—can divide in six minutes, beating even *E. coli,* but only by cheating, since whole nuclei depend on prefabricated ribosomes lurking in the vast resources of the newly fertilized fly egg.

The limiting factor on mutation rate is the finite size of the population in question and the deadly consequences of mutations hitting positions in the genome that are essential for life. Some viruses are highly mutable; for example, lentiviruses such as HIV have mutation rates as high as 0.1 percent per replication cycle. This is possible because their small genome of 9,000 base pairs would have on average one (or a few) serious changes and some will have zero. In addition, sharing of genomic material can occur between two adjacent viral genomes that are dysfunctional due to different mutations. In contrast, with 300,000 base pairs that matter per *E. coli* genome, and probably three times that for humans, the number of errors per base pair per division must be close to one per million (and can get better than one per billion).

How far could we push this if we could only mutate the nonessential or, better yet, the most likely to be useful bases, in order to succeed in our quest to turbo-charge the rate of evolution of organisms? If we had forty sites in the genome for which we would like to try out two possible variants in all possible combinations, then that would require a population at least $2^{40} = 10^{12}$ (a trillion) cells just to hold all of the combinations at once.

Those cells would all fit in a liter ($\approx$ quart). This would correspond to a mutation rate of forty genetic changes (rather than one) per genome per cell division. We could get away with fewer by spreading them over time or if the selection is additive. ("Additive" means that each change has some advantage and the order of change doesn't matter much.) So this additive scenario provides an alternative to exploring all 2^{40} special combinations at once. If we don't need to explore every combination exhaustively but want the highest mutation rate, that rate could be (theoretically) millions per cell per generation, depending on the efficiency of synthesizing and/or editing genomes (described below).

My gut feeling (by no means proven) is that, despite limitations of space and time, we humans can suddenly start to evolve thousands of times faster than during the impressive Cambrian era, and that we can direct this diversity toward our material needs instead of letting it occur randomly.

❧ ❧ ❧

The Real Point of Reading Genomes: Comparative and Synthetic Genomics

All these changes and innovations lie in the future. In the nearer term we will reap the benefits not only by manipulating genetic codes and mutating genomes but by reading them.

The object of the Human Genome Project (unsung predecessor to the well-known Personal Genome Project) was not, ironically, to read a real genome. Its goal (and final result) was to sequence a composite genome of several individuals, a veritable MixMaster blend of humanity. Furthermore, the genome that was actually sequenced was riddled with hundreds of gaps. While it was a historic milestone of science, it was nevertheless mostly symbolic—like the moon landing—and had relatively little value in practical, personal, or medical terms.

But supposing that we had sequenced a number of whole, intact, and genuine human genomes, the real point in reading them would be to compare them against each other and to mine biological widgets from them—

genetic sequences that performed specific, known, and useful functions. Discovering such sequences would extend our ability to change ourselves and the world because, essentially, we could copy the relevant genes and paste them into our own genomes, thereby acquiring those same useful functions or capacities.

A functional stretch of DNA defines a biological object that does something—and not always something good. There are genes that give us diseases and genes that protect us from diseases. Others give us special talents, control our height and weight, and so forth, with new gene discoveries being made all the time. In general, the more highly conserved (unchanged) the RNA or protein sequence, then the more valuable it is in the functions that it encodes, and the farther back in time it goes. The most highly conserved sequences of all are the components of protein synthesis, the ribosomal RNAs and tRNAs that transport amino acids from the cytoplasm of a cell to the ribosome that then strings the amino acids together into proteins. Even though these structures are hard to change evolutionarily, they can be changed via genome engineering in order to make multivirus-resistant organisms (Chapter 5), and by using mirror image amino acids, they can be changed to make multi-enzyme-resistant biology (Chapter 1).

Today it is possible to read genetic sequences directly into computers where we can store, copy, and alter them and finally insert them back into living cells. We can experiment with those cells and let them compete among themselves to evolve into useful cellular factories. This is a way of placing biology, and evolution, under human direction and control. The J. Craig Venter Institute spent $40 million constructing the first tiny bacterial genomes in 2010 without spelling out the reasons for doing so. So let's identify these reasons now, beginning with the reasons for making smaller viral genomes.

The first synthetic genome was made by Blight, Rice, and colleagues in 2000. They did this with little fanfare, basically burying the achievement in footnote 9 of their paper in *Science* ("cDNAs spanning 600 to 750 bases in length were assembled in a stepwise PCR assay with 10 to 12 gel-purified oligonucleotides [60 to 80 nucleotides (nt)] with unique complementary overlaps of 16 nt").

The authors had in fact synthesized the hepatitis C virus (HCV). This virus, which affects 170 million people, is the leading cause of liver transplantation. Its genome is about 9,600 bases long. The synthesis allowed researchers to make rapid changes to discover which of them improved or hurt their ability to grow the various strains in vitro (outside of the human body), which was a big deal at the time.

In 2002 Cello, Paul, and Wimmer synthesized the second genome, that of polio virus. This feat received more press coverage even though the genome was smaller (7,500 bases) and the amount of damage done to people per year was considerably less than that done by HCV (polio is nearly extinct). This heightened awareness was due in part to moving their achievement into the title: "Generation of Infectious Virus in the Absence of Natural Template." One rationale for the synthesis was to develop safer attenuated (weakened) vaccine strains.

In 2003 Hamilton Smith and coworkers synthesized the third genome, that of the phiX174 virus, in order to improve the speed of genome assembly from oligonucleotides. This exercise received even more attention, despite the fact that the genome was still smaller (5,386 bases) and didn't impact human health at all.

In 2008 the causative agent of SARS (severe acute respiratory syndrome), a coronavirus, was made from scratch in order to get access to the virus when the original researchers refused to share samples of it, perhaps under the mistaken impression that synthesis of the 30,000-base pair genome would provide an unacceptable barrier to their competitors. The researchers created the synthetic SARS virus anyway, for the purpose of investigating how the virus evolved, to establish where it came from, and to develop vaccines and other treatments for the disease it caused.

It turned out, however, that the synthetic versions of the SARS virus didn't work due to published sequencing errors. When the errors were discovered and corrected, the synthetic viral genome did work, and was infective both in cultured cells and in mice.

So, in summary, the reasons for synthesizing virus genomes are to better understand where they came from, how they evolved, and to assist in the development of drugs, vaccines, and associated therapeutics. Why did the

JCVI spend $40 million making a copy of a tiny bacterial genome? Some of this is attributable to the cost of research—full of dead ends and expensive discoveries. The other factor was that the core technologies used were first-generation technologies for reading and writing DNA, provided in the form of 1,000 base pair chunks by commercial vendors at about $0.50 per base pair plus similar costs for (semiautomated) assembly. So to redo a 1 million base pair genome would cost $1 million, while synthesizing a useful industrial microbe like *E. coli* would cost $12 million. But using second generation technologies for genome engineering (see below), not just one but a billion such genomes can cost less than $9,000. This is done by making many combinations of DNA snippets harvested from inexpensive yet complex DNA chips.

Most fundamentally, the object of synthesizing genomes is to create new organisms that we can experiment with and optimize for various narrow and targeted purposes, from the creation of new drugs and vaccines to biofuels, chemicals, and new materials.

In the quest to engineer genomes into existence, my lab mates and I have developed a technique called multiplex automated genome engineering (MAGE). The kernel of the technique is the idea of multiplexing, a term derived from communications theory and practice. It refers to the simultaneous transmission of several messages over a single communication channel, as for example through an optical fiber. In the context of molecular genetics, multiplexing refers to the process of inserting several small pieces of synthetic DNA into a genome at multiple sites, simultaneously. Doing this would make it possible to introduce as many as 10 million genetic modifications into a genome within a reasonable time period.

Here's how it works. To modify a genome, MAGE uses the smallest bits of DNA that are unique in the genome to be changed. The smallest unique pieces are theoretically around 12-mers for 5 million base pair bacterial genomes, since the number of ways that you can arrange four bases (A, C, G, T) into words twelve base pairs long is $4^{12} = 16$ million. In practice, the optimum seems to be 90-mers. We get these 90-mers into cells by means of a short, 2,500-volt electrical pulse. We chemically armor the ends of the 90-mer against enzyme damage in the cell, coat them with a protein that enhances pairing, and then let these pieces sneak into the gaps wherever the genome is replicating. Among other tricks, we use multiple 90-mers and multiple cycles. This collection of lab shortcuts and other techniques raises the chances of getting the desired genomic mutations from practically infinitesimal (one in a million) to very common (several per hour per cell). So, instead of assembling small DNAs into a huge piece of DNA, at great cost and with multiple errors, we use many small bits in parallel, a process that allows us to test several combinations for the purpose of seeing which work well and which are deleterious to the cell.

The other technique that greatly enables MAGE is selection, which is of course the key mechanism propelling Darwinian evolution. Selection in this context means rapidly separating out cells (or molecules) that have the more extreme values of the desired properties from all the rest. "Give me a fulcrum and I will move the world," from Archimedes, gets updated in the synthetic genomics lab to, "Give me a selection, and I will change the world."

We can easily select for leaving behind babies, since that is the best-tested selection paradigm of all time (tested by 10^{27} organisms for 3 billion years each). The players who leave behind the most surviving grandchildren tend to dominate subsequent history. This is of course a selection for all of the processes required for replication, not just the properties that we want to enhance. But frequently human engineers want to select for some property other than replication—for example, for overproducing milk in cows or producing insulin in bacteria, or artemisinin (an antimalarial precursor) in yeast.

Given that we want to engineer a whole genome, which is better? (1) Synthesize all of it, all at once, with possibly hundreds of design flaws to debug simultaneously? Or (2) make targeted changes combinatorially, dividing and conquering each bug swiftly and in parallel (and harnessing lab evolution). The Church lab voted strongly for (2).

Summing up, genome engineering already embraces much more than just genome copying, and much more than even the mature engineering disciplines (i.e., design and fabrication). Genome engineering has evolution on its side—not slow evolution but intelligently designed, fast evolution to create a range of special-purpose organisms that can produce desired substances.

Symbiosis, Morphology, and the Cambrian Diversity Explosion

So now let's suppose that we have created entire new genomes, and perhaps even some new single-cell organisms. The next step, obviously, is multicellularity.

One of the first signs of multicellularity, even before cells clump and differentiate, is sex. Exchanging genetic material allows for highly parallel discovery of new genomic sections. For example, if it takes a day for a bacterial cell to replicate in the wild, it will take thirty days for a mutant cell to replace a population of about a billion cells, and it will take 100,000 years to change 20 percent of a bacterial genome (a working definition of a new species). This assumes that selection pressure is fairly steady in one direction and that there is a constant stream of beneficial mutations. In practice a significant fraction of the evolutionary steps toward that 20 percent are neutral, neither enhancing nor harming the organism, and the number of setbacks are such that a million years is not unusual for the full appearance of new species. If new favorable mutations can combine by means of sex, then the time spent sweeping away the competing population is saved (keeping in mind that sex can separate favorable alternative forms of a gene as well as bring them together).

To drive all this toward its limits, we are using a second lab technique— conjugative assembly genome engineering (CAGE)—selectively mating two bacterial strains in order to combine parts of their genomes. If MAGE is a means of breaking up a long piece of DNA and making small changes in each part, CAGE is a means of putting the changed parts together again. This allows us to divide and conquer the *E. coli* genome, for example, into thirty-two pieces. We first apply the MAGE technique, focusing on only 140 kbp in each of thirty-two separate strains in parallel. We then mate strain 1 and 2, bringing together adjacent sectors in the genomes of the strains. Similar matings bring together sixteen even-numbered sectors with the adjacent sixteen odd-numbered sectors, thereby reducing the number of strains by a factor of two. Then we repeat the process to get down to eight strains, then four, then two, and then finally one bacterial strain.

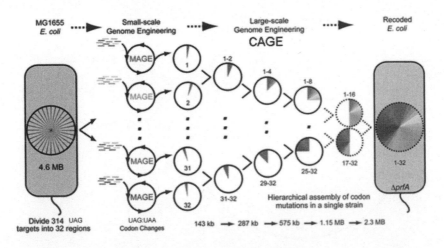

Figure 3.3 Divide and conquer with MAGE and CAGE. For example, to change all UAG stop codon targets (322 total in the *E. coli* genome) into the functionally identical codon UAA, we divide this task into thirty-two sectors of the genome pie (far left) and design thirty-two pools of ten oligos each (second column) that can change ten codons in each sector (third column). By selective mating we merge adjacent pairs of sectors (e.g., 1 and 2) resulting in sixteen bacterial strains, each with a double-size sector (fourth column) in the genome of that strain. This process of merging adjacent sectors continues to eight, then four, then two (columns 5 to 7); the final mating results in one strain that has all of the UAG codons changed to UAA, as was planned.

A second advantage of multicellularity is cross-feeding, where one cell type can focus on one type of food chemistry and can barter its specialty food for a different food from an adjacent cell. You can divide your genome over two or more cell types and hence save space, modularize and parallelize selection, and quickly benefit from sharing the multitude of tasks.

A third advantage of multicellularity is shape selection. Single cells have some diversity in morphology, but their limit becomes rapidly evident. For example, making a cell that is as hard as a tooth doesn't automatically confer chewing ability, since that requires structures larger and more complex than a cell. How to get from the straight (one-dimensional) DNA to complex three-dimensional dynamic shapes is still mysterious. But, as always, synthesis offers us both a path to discovery and a rigorous assessment of progress toward understanding the processes involved.

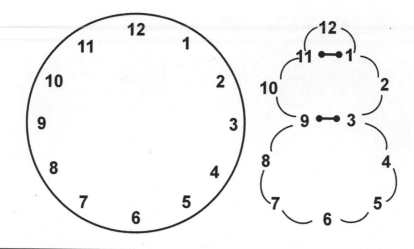

Figure 3.4 Basic DNA origami. One long (single-stranded) DNA circle is represented as a twelve-hour clock face on the left. The addition of two (single-stranded) staple oligos connect (via base-pairing each segment of the circle) 11 to 1 and 9 to 3. This results in a shape resembling a Dali-esque odalisque with a double-stranded collar and belt.

Between 1977 and 1981, based on the first folded RNA structure (tRNA, Figure 3.1), Ned Seeman and I invented ways to design and establish morphology from the basic base pairing rules of RNA and DNA (G with C, and A with U/T, Figure 1.4). We can now design and build atomically precise shapes swiftly and with generally high yield. We call these shapes DNA nanostructures (from Ned Seeman and William Shih), or DNA origami (from Paul Rothemund).

As a further example, 170 DNA staples (40-mers) bind to precise locations around a 7,000 base long single-stranded circle with half of each 40-mer staple binding one place and the other half binding somewhere else (designed by caDNAno, a computer-aided design process).

Shawn Douglas, Ido Bachelet, and I at the Wyss Institute for Biologically Inspired Engineering at Harvard have used this technique to construct nano robots—essentially cages made of DNA that hold cancer-killing antibodies. These nano cages open up and release their cargo of antibodies only when they touch cancer cells. The rules for defining the shape of pro-

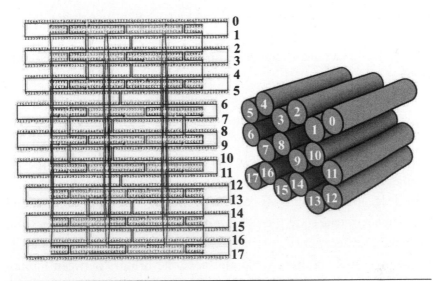

Figure 3.5 3-D DNA origami. One long DNA circle (represented as mostly horizontal lines plus far left and far right loops) is guided to fold into a specific 3-D log pile of eighteen double helices. The short, linear staples (mostly vertical lines plus internal loops) connect specific distant points on the circle.

tein structures are considerably harder than those for designing DNA structures (the awesome base pairs of Figure 1.4). The rules for designing cell and multicellular shapes are harder yet (unless, of course, we've seen them before).

Despite the three advantages listed above, multicellularity tends to come with a hard consequence—you give up your immortality. Most single-cell organism species seem to live indefinitely, its members still evolving, traceable as a continuous genetic lineage (especially in the highly conserved ribosomal RNA genes) back to the dawn of time. Selection in species with diverse cell types comes along with planned obsolescence and a circle of life progression—from egg to chicken to egg. As single-cell organisms replicate by simple cell division, lessons learned (mutations and environmental responses) just prior to replication are often retained in the daughter cells. But with multicellular organisms, the larger they are, the more phases they have to pass through in going from egg to adult. In species that learn, a lifetime of experience is lost at death. This loss can be partially compensated for by inheritance of cultural (non-DNA) artifacts, such as in the *Cyanistes* birds in the 1960s that taught each other how to open foil-topped British milk bottles.

In any case, the advantages and disadvantages of multicellularity may no longer apply to humans because of the huge impact of cultural inheritance and the prescient design of our technological culture. The advantages of multicellularity (reassortment of genes, division of labor and shapes) are now affected by markets and software. The downside of knowledge lost at death is offset by vast libraries and education, but a full education can take sixty years. So we'd like to either speed up education or slow down decay once a person is educated. The former might be made possible by making more rapid and accurate bridges between human knowledge and computer knowledge, either by the optimal use of existing senses or by means of multi-electrode neuronal input and output. Retarding the processes of decay, by contrast, might be made possible by discovering why some animals live long and vigorous lives while others die quickly (ranging from 3 days to 400 years; see Chapter 9). To study this phenomenon, we'd like to have two species that are closely related but have radically different life spans. As for example . . .

The Naked Mole Rat and Longevity

As a physical specimen, the naked mole rat is the stuff of nightmares. With its saggy pink skin, piglike nose, spindly legs, tiny, almost vestigial eyes, and mere holes for ears, it looks like the ultimate misbegotten animal.

This highly unusual specimen isn't even a rat, strictly speaking, nor is it a mole. It belongs to a genus (*Heterocephalus*) that has no other known members. It lives up to its name in being "naked," for it is almost hairless. The mole rat spends virtually its entire life underground in total darkness, in a complex maze of subterranean channels and tunnels whose cumulative length can add up to two miles or more. This is remarkable because these are small animals, generally about the length of a human finger, and weigh little more than a mouse. Native to the hot grassland regions of Kenya, Ethiopia, and Somalia, the mole rat does not drink water (or anything else!). It can run backward and forward equally fast. Because its skin lacks a key neurotransmitter that in mammals is responsible for transmitting pain signals, the naked mole rat can feel no skin pain.

As if it's not already distinctive enough, the mole rat has a social structure that is almost unique among mammals. The species is "eusocial," meaning that its colonies are organized like those of ants or bees, with the members existing in strict hierarchical castes. At the top is a queen who breeds with only a few select males. Next down are the soldiers, who defend the colony against predatory snakes or foreign invader rats. At the bottom of the social scale are the workers, who forage for food, mainly roots and tubers.

But however odd their appearance and behavior—one observer has called them "fauna incognita"—naked mole rats possesses two additional characteristics that make them of special interest to biologists. The first is that they are the world's longest-lived rodent. Whereas the house mouse, for example, has an average life span of two or three years, the naked mole rat can live for twenty-five years or more (the current record is 28.3 years). The second is that they are extremely resistant to cancer. Indeed, cancer has never been detected among these animals.

These two facts illustrate the importance of the new science of comparative genomics. Genomics in general relies on our ability to read, or sequence, the DNA of a given organism. One goal of sequencing the human genome is to identify genes that play a role in disease (see Chapter 9). But reading genomes has another and equally important objective, which is to find useful biological widgets in other organisms—special-purpose apps, as it were. Comparative genomics, the study of how similar genes function across different species, will allow us to locate genetic structures that confer distinct advantages on certain classes of organisms. The long life span of the naked mole rat is one example. Its longevity is a trait that must be rooted somewhere in its genetic makeup. If we can find the gene—or more likely the combination of genes—that gives such great longevity to the animal, this will be a genomic component that we can exploit to our benefit, and perhaps even import to the human genome.

In humans, old age is a factor in ailments such as heart disease, type 2 diabetes, cancer, and neurodegenerative diseases including Parkinson's and Alzheimer's. But nobody knows why some organisms have fleeting life spans while others live for a century or more. There is at least a forty-fold variation in maximum longevity among mammals. The white-faced capuchin monkey has a life span of over fifty years. Humans can live for over one hundred years. And then there's the case of the bowhead whale: with an estimated life span of over two hundred years, the bowhead whale is the only mammal known to outlive human beings, and is possibly the longest-lived mammal on earth.

Still, it's a mystery why different species that share a similar body plan, biochemistry, and physiology nevertheless age at such different rates. Comparative genomics may help us solve the riddle. Sequencing the genomes of these long-lived mammalian species may reveal a set of homologous (similar or shared) genes responsible for their extended life spans. Discovering these genetic structures would provide us with insights into the mechanisms of aging and of age-related human diseases, and this in turn will lead to better diagnoses and treatments.

In 2007 a group of researchers, including myself and my colleague Joao Pedro Magalhaes at the University of Liverpool, supported by seventy-nine scientists from other institutions, formally proposed sequencing the genomes of the naked mole rat, the capuchin monkey, and the bowhead whale to the National Human Genome Research Institute (NHGRI). This initial proposal was rejected, essentially on the grounds that it is a long way from knowing the respective sequences to understanding exactly how the genetic structures in question function to lengthen life span. This may be true, but knowing the sequences would nevertheless be a genuine first step toward making progress in solving the problem.

A year later, the same group, supported by the same seventy-nine scientists, submitted a second proposal, this time to sequence the genome of the naked mole rat alone. Not only does the mole rat have exceptional longevity, but it also seems to be immune to cancer. But this proposal too was rejected, on the grounds that identifying the precise complex of genetic structures that underpin the longevity of the animal would be difficult.

Subsequently we started the sequencing ourselves, on a shoestring budget, and cadging spare sequencing capacity from friends. The first data for the naked mole rat genome and RNA started flowing in the spring of 2011. By summer Magalhaes and his team at Liverpool had made the first draft sequence and put it online. As of this writing we are still in the midst of interpreting the sequence (the hard part), as well as planning follow-up experiments.

The list of genes that scientists have discovered grows by the day: there are genes for cystic fibrosis, skin cancer, lung cancer, and on and on. Other genes control height, weight, and a host of other traits. Comparative genomics has been a huge help in finding those structures, and genome engineering will make it possible to incorporate them, gradually or swiftly, into the human genome.

Undoubtedly, some people will object to modifying the human genome in this way on a variety of grounds: moral, philosophical, political, religious, aesthetic—and let's not forget just plain emotional grounds (or even no grounds whatsoever). Objections to new technologies (see the technology prohibition plot in the Epilogue) typically peak as the technology is poised to spread among early adopters but doesn't yet work well. Then, once the technical bugs are ironed out, the moral high ground can invert. For example, in vitro fertilization was considered unnatural and risky at first, but eventually withholding access to the procedure from infertile couples became unacceptable. Vaccines have had numerous periods of bad press since Pasteur's rabies tests, and even Edward Jenner's first trials of smallpox vaccination were ridiculed by some British cartoonists. But whenever an outbreak occurs, especially after a public campaign that reduces local vaccinations, the popularity of vaccines mysteriously increases. Sometimes people invoke the precautionary principle of "do no harm," but in some cases doing nothing is harmful in and of itself.

So, with respect to human longevity, how many of us really want the status quo prolonged? Or how about the longevity of our pets? Dog owners and cat lovers typically outlive several successive pets, experiencing wrenching partings with their animals every time one of them dies a natural death or is euthanized. Wouldn't it be better if your pet could be made to live as long as you (or at least double its normal life span), all the while remaining in good health?

And then there's this question: Is anyone actually in favor of aging? Some worry that a widespread increase in human life span could cause overpopulation. But "overpopulation" is a relative concept, and by some people's reckoning the world is already overpopulated and indeed it was overpopulated decades ago (see Chapter 9). As our human population is dramatically shifting to cities, the average family size is falling to below the replacement level of 2.1 children per couple. Further, it is a well-established trend that as people become wealthier, they have fewer children. Finally, as we have seen, there is also the option of getting some of the population off the planet.

Lysenko and Eugenics: The Future of Cultural Evolution

The ideas surrounding eugenics and Lysenkoism are nearly synonymous with bad science—worse than merely mediocre science because of their huge and adverse political and economic consequences. Here we will re-examine these ideas from a radical new perspective to see if, against all expectation, some value can come from these most unlikely quarters.

Trofim Lysenko was a Soviet biologist who accepted the Lamarckian theory of the inheritance of acquired characteristics, which opposed the Mendelian theory's view that inherited characteristics are inborn and not affected by the environment. In 1940 Lysenko became the director of the Soviet Institute of Genetics, and, with Stalin's backing, applied his version of Lamarck's 1822 theory to agriculture. Lysenko's one big idea was vernalization—pretreating seeds with cold and moisture so that they would sprout and grow earlier in the spring than untreated seeds. He further held that vernalized seeds would give rise to plants whose seeds were also vernalized because they had acquired that characteristic through inheritance. (Which they did not in fact do.)

Although this practice did not bode well for Soviet agricultural production, Lysenko's ideas, collectively termed Lysenkoism, nonetheless became the official agricultural dogma of the USSR. Those who opposed it were persecuted, imprisoned, and sometimes even killed.

In the 1960s, Andrei Sakharov and other Soviet physicists finally precipitated the fall of Lysenkoism, blaming it for the "shameful backwardness of Soviet biology and of genetics in particular . . . and for the defamation, firing, arrest, even death, of many genuine scientists."

At the opposite (yet equally discredited) end of the genetic theory spectrum was the Galtonian eugenic movement, which from 1883 onward grew in popularity in many countries (including the United States, the United Kingdom, and Germany). In its extreme form, eugenics propounded the forced sterilization of various "undesirables," and this was perpetuated despite the 1948 Universal Declaration of Human Rights, which proclaimed that "men and women of full age, without any limitation due to race, nationality or religion, have the right to marry and to found a family." In fact, forced sterilization persisted into the 1970s in Sweden and Canada.

The conventional wisdom regarding these two pseudoscientific movements is that Lysenkoism overestimated the impact of environmental influences while eugenics overestimated the role of genetics. But an interesting and radical alternative interpretation is that both theories *underestimated* and in fact hobbled both of these powerful forces: they tried to apply genetics on a grand economic and human scale without being able to directly recode the genome.

One form of scientific blindness occurs, as above, when a theory displays exceptional political, faith-based, or intuitive appeal. But another source of blindness arises when we rebound from catastrophic failures of pseudoscience (or science). For example, Lysenko's spectacular failure in his attempts to apply a Lamarckian view of evolution can blind us to the ways in which we do in fact inherit acquired characteristics, for example, through epigenetics. The grandchildren of those who survived the 1944 Dutch hunger winter had smaller than average birth weights.

Our children already inherit our computers and cars as surely as they inherit our brains and brawn. Indeed, we have inherited acquisitions ever since we developed tools and domesticated animals. But now this form of inheritance has become increasingly dominant (over genetic inheritance) and rapidly exponential. Even genetic inheritance could become genuinely Lamarckian if we became as confident and adept in applying our synthetic biotechnologies as we have been in the application of our inorganic technologies. We have always applied genetics in a weak sense, and in general unwittingly, at the individual family level, by marrying whomever we want, and for the genetically based characteristics we see embodied in those we choose.

Many have speculated that human evolution has stopped. But we are well into an unprecedented new phase of evolution in which we must generalize beyond our DNA-centric worldview. Evolution can accelerate from geologic speed to Internet speed—still employing the processes of random mutation and selection, but also by the use of nonrandom, intelligently designed genomes, and by use of lab selections, which makes the process even faster. We are losing species—not just by extinction but by merger. The species barriers separating humans, bacteria, and plants were breached occasionally over vast evolutionary time frames through horizontal gene transfer. But today those high barriers have vanished. Genes for bacterial insecticides and for herbicide resistance are permanently integrated into current crop genomes. Even the barrier between humans and machines is porous—think of pacemakers, artificial hearts, hearing aids, and cochlear implants. Between humans and other animals we have xenotransplantation, the use of pig heart valves in human hearts, and so on.

One objective of eugenics was to improve the intelligence of the general population. But this is happening already, without the use of Galtonian-style eugenics measures. Consider the "Flynn effect," the observation that standardized test scores for general intelligence aptitude have been increasing since 1932, when the first of these tests was introduced. Various explanations have been offered for this, including better nutrition, greater public exposure to testing, an overall increase of stimulation and information by means of television and the Web, the increased use of "short-hand abstractions" (scientific terms that have been exported to general usage, such as "placebo effect" and "random sample"), lowered infectious load, and the crossbreeding of formerly inbred populations (heterosis), among other things.

Although barely noted and not contributing to the Flynn effect (as yet), the first permitted use of calculators in SAT tests probably marked a major milestone in the man-machine merger. How many of us have participated in conversations that are semidiscreetly augmented by Google or text messaging? Even without invoking artificial intelligence, such commonplace enhancements of our decision making amount to nongenetic ways of augmenting intelligence. In parallel, incremental improvements in current blood stem cell transplantation will make us more confident in the safety and efficacy of adult stem cell genome engineering. Such clinical genetic interventions will not be usefully lumped together with eugenics, especially if they are confined to somatic cells and not germ line cells, if they are done voluntarily by individuals and families and not at the behest of governments, and if they are diverse and not monochrome.

The concepts of maximizing evolution by means of population size, speed of mutation, replication, selection, and recombination apply here too, although their effects are harder to estimate, especially as we consider the way in which our cultural and technological artifacts increasingly become part of our evolutionary life. With "generation" times (i.e., the time between applying selection to cultural variations "generated" on the Internet) possibly in the nanosecond range (rather than the seven minute world-record minimum for replicating a living cell), and with 10^{18} bytes "selected" per day, this form of evolution starts to get interesting. Every day the composition of which 10^{18} bytes are sent over the Internet is selected by economic, intellectual, and entertainment (Darwinian) selective forces. Our computer-aided evolution can enable us to inherit acquired traits, after Lamarck, while gene pools of accelerated evolution will be subject to Galtonian market pressures. In Chapter 9, for example, we will see how Nic Volker and Timothy Ray Brown have become living testimonials to the power of stem cell transplants to change body genetics (to eliminate intestinal problems, leukemia, and AIDS), and how their newly "acquired" state will likely spread to many other patients by viral word of mouth.

What limits the number of computer replication and selection operations? Energy. Right now computers accomplish 10^9 operations per Joule while DNA replication is far more efficient at about $2x10^{19}$ operations per

Joule. So, biologically inspired improvements in computer efficiency might lie in the near future. For example, the encoding digital information in DNA (text and images, using the scheme A:oo, C:o1, G:1o, T:11, as described more fully in Chapter 8), in addition to being potentially a billion times more compact, less expensive, and longer-lived than paper or CD/Blu-ray disks, could be more efficient to manufacture and search.

The second part of the energy equation is the cost of acquiring the energy from a renewable source. Read on.

CHAPTER 4

-360 MYR, CARBONIFEROUS
"The Best Substitute for Petroleum Is Petroleum"

On Sunday, February 24, 2008, a Virgin Atlantic Boeing 747 flew from London to Amsterdam with a blend of 20 percent biofuel and 80 percent standard Jet-A in one of its fuel tanks. It was a short trip, only 221 miles, and lasted only 70 minutes; moreover, as a demonstration flight the plane carried no paying passengers. Nevertheless, this was the first flight by a commercial airliner that was powered in part by biofuel (in this case a mix of coconut and babassu nut oils). Richard Branson, CEO of Virgin Atlantic Airlines as well as Virgin Fuels, called the flight "historic."

It was the first of a series of proof-of-concept test flights. Later that year, in December, an Air New Zealand Boeing 747 made a two-hour demonstration flight from Auckland International Airport. With one of its four Rolls Royce engines powered by a fifty-fifty blend of jatropha oil and standard jet fuel, the plane climbed to its normal cruising altitude of 35,000 feet. The pilot shut down the bio-fueled engine, restarted it, and then performed a number of other exercises before making a routine landing. "We undertook a range of tests on the ground and in flight with the jatropha biofuel performing well through both the fuel system and engine," said the carrier's chief pilot, David Morgan.

About two weeks later, on January 7, 2009, a Continental Airlines Boeing 737 departed from Bush International Airport in Houston, Texas, with one of its fuel tanks containing the most exotic brew yet: a mixture of 50 percent conventional jet fuel, 47.5 percent jatropha oil, and, something new, 2.5 percent algae-derived biofuel. For two hours, the plane flew a series of maneuvers over the Gulf of Mexico, including a midair engine shutdown and successful restart, after which it landed without incident. "This is really a kind of landmark," pilot Rich Jankowski said afterward.

These three flights were a window onto the coming era of biofuels, for by the summer of 2011 at least six airlines, including KLM, Lufthansa, and Finnair, had used biofuels on commercial flights carrying paying passengers. With the global fuel market operating at a trillion dollar level, diminishing oil reserves in the ground, much of it controlled by unstable or unfriendly political regimes, and a global warming crisis caused in large part by carbon emissions from the burning of fossil fuels (not to mention the human, economic, and environmental costs of disasters such as British Petroleum's Deepwater Horizon oil rig blowout in the Gulf of Mexico in April 2010), the idea of renewable energy sources exerted a powerful appeal.

In fact, the whole idea of biofuels, especially the vision of having microbes such as cyanobacteria "grow" petroleum for you, seemed to be surrounded by a halo effect, a magical radiance. After all, it was like getting something for nothing, or almost. The idea was that we're going to leave the fossil fuels in the ground and put microorganisms to work for us, producing whatever alternative fuels we need. As a side benefit, those same microbes would clean up the atmosphere and help save the world in the process.

Who could resist such a dream? Not many. As the twenty-first century began, airlines, automakers, national governments (especially the military components thereof), and even the oil companies themselves were all jumping on the biofuels bandwagon, and looking to replace, or at least supplement, fossil fuels with fuels that are *grown*, like maple syrup, tomato juice, and coconut milk (or for that matter, cow's milk).

By 2010 there were more than two hundred companies in the United States, including one located on the Southern Ute Indian Reservation in Ignacio, Colorado, plus dozens more abroad, competing for the potential fortunes to be made from microbes that produce fuels. But there was more to biopetroleum than biofuels. In 2011 retail giant Walmart began selling a bio-based motor oil, G-Oil, which was being advertised in car magazines as "green bio-based full synthetic motor oil." The ads ran under a banner headline, "Change your oil, change the world," while a text at the bottom announced that the oil was "Grown and made in the USA." What the ads failed to mention was that the oil was not made by special-purpose and efficient microbes, but was based on some of the least efficient biosources of all—rendered beef, pork, and chicken fat. (How an oil made from animal fat could be "fully synthetic" was not explained.) Nevertheless, the fact that such a product was now being marketed by a major retailer (and used as a lubricant in Mazdaspeed Formula One racing cars) showed that the idea of bio-based petroleum had suddenly come of age. All at once, it was glamorous.

Of course there were a few minor problems with this otherwise rosy scenario. One of the earliest start-ups to enter the algae-to-biofuels game was GreenFuel Technologies Corporation. Founded in 2001 and based in Cambridge, Massachusetts, it planned to produce vast quantities of algae using CO_2 smokestack emissions from power plants, and then use the resulting masses of algae to make biodiesel, among other things. The company christened this process as its proprietary Emissions to Biofuels technology, and raised $70 million in private funding. By 2005 the company had established a working bioreactor pilot plant in Arizona. But two years later the pilot plant was producing more algae than it could convert into fuel. Growing the algae, the company discovered, was the easy part. The hard part was getting the algae to make petroleum, especially in a cost-effective way. Further compounding its difficulties, the company did not perform any genetic engineering on the microbe it was using. In 2009, having been blindsided by the fine print of microbiology, GreenFuel filed for bankruptcy.

California-based Solazyme provided another cautionary tale. Unlike GreenFuel Technologies, Solazyme did not go out of business. To the contrary, the company was wildly successful, announcing in 2010 that "we delivered over 80,000 liters (21,000 gallons) of algal-derived marine diesel and jet fuel to the U.S. Navy, constituting the world's largest delivery of 100% microbial-derived, non-ethanol biofuel." What the company did not reveal, although the *Marine Corps Times* did, was the price per gallon of this otherwise auspicious and forward-looking new substance: $424.

No consumer in a calm and sober frame of mind would pay anything like that kind of money for a gallon of gas. But the military, which was famous for its thousand dollar toilet seats and wrenches, had a totally different mind-set when it came to spending. But if the military was your major customer for algal-derived biofuels, then you had to wonder about the real-world viability of the product. Was it really anything more than a pipe dream?

The Virgin Atlantic test flight was powered in part by coconut oil. But one critic (Jeff Gazzard of the Aviation Environment Federation, based in the UK) quoted an estimate published in *Petroleum Week* that if the flight had been made entirely on coconut oil, it would have consumed 3 million coconuts. By any standard, that's a lot of coconuts.

Apparently, though, *Petroleum Week* made a slight miscalculation. According to Wikipedia, a thousand mature coconuts yield approximately 70 liters of coconut oil, which is to say that fourteen coconuts make one liter of the stuff. And according to the Boeing.com website, the 747 family of aircraft has an average fuel mileage consumption rate of nineteen liters per mile. So a flight of 221 miles would require 221 x 19 = 4,199 liters of coconut oil. And at the rate of fourteen coconuts per liter, the flight would require only 4,199 x 14 = 59,985 coconuts.

But that's *still* a lot of coconuts. And people eat coconuts.

By contrast, both the Air New Zealand and Continental Airlines demonstration flights were powered in part by jatropha oil, which, because it's made from a nonfood crop, at least had the advantage of avoiding the food-for-fuel problem (unless land normally used for food is displaced to grow the jatropha plants). Nevertheless, the global aviation industry currently burns through about 240 million tons of jet fuel per year, and one industry journal calculated that producing that amount of fuel from jatropha alone would require planting a land area that was twice the size of France.

So maybe biofuels really are the fuel of the future—and always will be.

﷽   ﷽   ﷽

If time travel ever becomes possible, the Carboniferous period, which lasted for some 74 million years, from about 360 million to 286 million years ago, would be a good era to avoid. Huge insects crawled, crept, and flitted across the earth. Two of them were the largest known insects of all time, the centipede *Arthropleura*, which grew to a length of more than eight feet, and the giant dragonfly *Meganeura*, which had a wingspan of some two and a half feet. These enormous dimensions were possible because at that time oxygen made up 35 percent of total air volume (rather than our current wimpy 21 percent).

Other flora and fauna of the period included several tree species, seed-bearing plants, mosses, and fungi; sponges, corals, sharks, and other fish in the oceans and lakes; and on land, numerous varieties of four-footed animals including lizards and reptiles. The emergence of reptiles was a product of a major evolutionary innovation that took place during the Carboniferous, the development of the amniote egg, whose outer shell and inner membrane protected the embryo inside it from drying up.

It was also in the Carboniferous when the source of the planet's current greatest scourge and atmospheric nightmare originated, fossil fuels. Fields of coal, petroleum, and natural gas are thought to have been laid down during this time. The term "Carboniferous" means carbon-bearing, "carbon" being the Latin word for coal. The earth's coal beds were the product of the vast swamps and forests that covered most land masses. The dead plants washed into the sea and sank into the mud, where tectonic forces subjected them to great pressure and heating that, across millions of years, converted them into coal.

Petroleum is thought to have originated in a similar manner, but not from the bodies of dead dinosaurs, as many have imagined. That idea is arguably a relic of the Sinclair Oil Company's decades of advertising using the image of "Dino," its trademark green Brontosaurus, which was an instantly recognizable corporate symbol. Unfortunately for Sinclair, the dinosaurs were creatures of the Mesozoic, which followed the end of the Carboniferous period by some 35 million years. Which means that the oil was already in the ground and mellowing long before dinosaurs even existed.

But if the dinosaurs didn't give us petroleum, then what did? For a long time the answer to this was controversial, and in his 1979 book *Energy from Heaven and Earth*, the physicist Edward Teller (the so-called father of the H-bomb) studied the question and reported, "I have gone to the best geologists and the best petroleum researchers, and I can give you the authoritative answer: no one knows." Indeed, when he wrote those words, matters were not so clear as they have subsequently become. Even today, however, there is a generally accepted "official" theory of petroleum's origin, as well as a competing alternative theory.

According to the orthodox view, petroleum was formed from the decomposition of organisms that settled to the sea bottom. These organisms may have been phytoplankton (microscopic plants floating in water), zooplankton (small animals), or both. Petroleum supposedly arose from them in the way that decaying plants gave rise to coal. Over the eons, the zooplankton and/or the phytoplankton became buried under increasing amounts of rock. This rock buildup created pressures that were great enough, as well as temperatures that were high enough, to cause the organic material remains to go through a series of chemical reactions that transformed them into the hydrocarbons of crude oil.

This is the so-called biogenic theory of petroleum's origin. Supporting it is the fact that many oil reserves are found underwater, which explains the widespread practice of offshore drilling. In addition, oil contains biomarkers, molecular structures associated with biological and plant organisms. If oil didn't arise from these organisms, the argument goes, then how did the biomarkers get there? While these points would seem to settle the issue, the alternative theory of the origin of petroleum, the abiogenic theory, offers a different explanation of both the biomarkers and the underwater location of oil reserves.

The abiogenic theory holds that the ingredients that make up petroleum were brought here at the very beginning, as part of the earth's formation. They came with the planet, as it were, and were here all the time, hidden beneath the earth's crust. If the abiogenic theory is correct, then petroleum and natural gas are not fossil fuels at all because they did not arise from fossils. As an account of petroleum's origin, the abiogenic theory at least has the virtue of simplicity.

Also in favor of the abiogenic view, whose most vocal proponent in the United States was the Cornell astrophysicist Thomas Gold, are the facts that carbon is the fourth most abundant element in the universe, and that it exists predominantly in the form of hydrocarbons. In our own solar system, furthermore, huge amounts of methane (a hydrocarbon) are found in the atmospheres of Jupiter, Saturn, Uranus, and Neptune. But if the universe in general, and the solar system in particular, is teeming with hydrocarbons, then why do we need any special act of creation of petroleum based on the decomposition of organisms?

The abiotic theory claims that petroleum's biomarkers got there as a result of contamination. A distinct class of organisms, the so-called extremophiles, exist at depths where petroleum drilling takes place, and the oil becomes commingled with biological materials (i.e., the biomarkers) during the drilling process. (The underwater location of oil reserves is explained by the fact that about 70 percent the earth's surface is water.)

The simplicity of the abiogenic theory makes it intuitively plausible, but that apparent straightforwardness masks several deeper layers of complexity. (There is even a duplex theory, propounded by the British synthetic organic chemist Sir Robert Robinson, that holds that young, or juvenile, oils arose biogenetically whereas older, ancient oils did so abiogenically.)

The dominant theory today, however, and the one held by most geologists, is that oil arose from microorganisms. In 2006 the British chemist Geoffrey Glasby published an exhaustively researched study of the abiogenic theory and concluded on the basis of several technical factors that the theory doesn't work. Even he, however, conceded that "the relative roles of bacteria in the formation of petroleum, in degrading the initially formed petroleum to heavy oil and in supplying biomarkers to hydrocarbon deposits are still not fully understood."

Whether or not petroleum came from microorganisms or was here from the very beginning, there are more sources of diesel oil, gasoline, and jet fuel than the crude oil pumped from the ground. Living organisms can be coaxed to create more of any desired "fossil fuel." The question is whether they can do so at a competitive cost.

ϡ ϡ ϡ

Biofuels are nothing new. Wood was the first biofuel, used for heating and cooking. During the nineteenth century, whale oil was used for lighting (as well as for soap making). At the 1900 World's Fair (Exposition Universalle) in Paris, Rudolf Diesel ran his eponymous engine on peanut oil. He was an early advocate of biofuels, saying at one point that oils derived from vegetables and other plants "make it certain that motor power can still be produced from the heat of the sun, which is always available for agricultural purposes, even when all our natural stores of solid and liquid fuels are exhausted." An optimist. Henry Ford built cars that ran on ethanol, babassu nut oil, and soybean oil.

Those same oils, plus many others, have been proposed by biofuel proponents as substitutes for petrochemicals or as feedstocks from which biofuels can be made. The list of potential biofuel sources is rather grand, and includes corn, peanuts, coconuts, babassu nuts, wheat, sugar cane, sugar beets, molasses, cassava chips, canola oil, castor oil, cottonseed oil, pumpkin seed, beechnut, chestnut, lupine seed, poppy seed, rapeseed, linseed, peas, olives, sunflowers, palm, fish, animal fat, soybeans, jatropha, mahua, mustard, sweet sorghum, camelina, switch grass, Miscanthus grass, straw, seaweed, pine chips, plus miscellaneous organic waste, hemp, and shea butter, among other things.

However, there are good and sufficient reasons why, even in an age in which biofuels already enjoy a cachet, conventional oil continues to win by a landslide in the marketplace. Consider it against ethanol, for example, which is a widely used biofuel. In favor of ethanol are two main facts: one, the corn from which it is made is a fast-growing, cheap crop that is planted and harvested all over the globe, making it a renewable resource par excellence. Second, ethanol is a net zero emissions fuel because the CO_2 produced when a car burns it is offset by the CO_2 sequestered by the corn's growth process.

Nevertheless, corn-based ethanol is hardly the ideal fossil fuel replacement. For one thing, it has a lower energy density than gasoline, which means that a gallon of ethanol produces less energy than a gallon of gasoline—some 35 percent less, which means that you have to burn more of it to get the same amount of energy out of it. This explains why a biofuels company that I co-founded, LS9, which produces renewable petroleum by means of designer microbes, takes as its corporate slogan, "The best replacement for petroleum is petroleum."

Second, making ethanol from corn removes corn from the global food market, and also from people's mouths. In fact, the process of ethanol production gives rise to a whole train of unintended consequences and ripple effects, effects that bring to mind the parable by Frédéric Bastiat, the French economist who in 1850 wrote *Ce qu'on voit et ce qu'on ne voit pas* ("that which is seen and that which is unseen"). In the parable, a shopkeeper's son accidentally breaks the store's large window, an act that is universally seen as a boon to the glazier who replaces it for six francs. (And window breakage more generally is popularly regarded as a windfall to glaziers.)

But that is only what is *seen*, says Bastiat. "It is not seen that as our shopkeeper has spent his six francs upon one thing, he cannot spend them upon another. It is not seen that if he had not had a window to replace, he would, perhaps, have replaced his old shoes, or added another book to his library. In short, he would have employed his six francs in some way, which this accident has prevented."

In 2007 a group of researchers studied the unseen consequences of a large-scale replacement of fossil fuels by corn-based biofuels such as ethanol. In "The Ripple Effect: Biofuels, Food Security, and the Environment," the authors argued that a switch to such fuels would have a range of adverse and unintended effects: increased food prices, coupled with reduced availability, as corn was diverted to biofuel production. (And because corn is used for cattle feed, higher corn prices would raise the price of beef.) In addition, some lands formerly used for agricultural purposes would now be diverted toward biofuel production, thus also raising the price of farmland. That price rise would in turn drive farmers to boost crop yields by greater use of fertilizers, with the result of increased ground-water pollution and higher atmospheric levels of nitrous oxide, a greenhouse gas. And so on.

During the second half of 2010, the price of corn rose 73 percent in the United States, in large part due to the use of corn for ethanol production. Hence the appeal of biofuels that are not made out of foodstuffs but are rather synthesized—grown—by microbes.

$$\text{\&}\quad\text{\&}\quad\text{\&}$$

When the Continental Airlines Boeing 737–800 made its biofuels demonstration flight in January 2009, a fraction of its fuel component had been made by algae cultivated by Sapphire Energy of San Diego, California. Sapphire, along with Solazyme, was a biofuel success story. Financed by venture capital firms owned by Bill Gates and by the Rockefeller family, the company had produced its jet fuel in a field of open ponds on a 100-acre algae farm near Las Cruces, New Mexico, using nothing but natural (non-genetically modified) algae, CO_2, nonpotable water, and sunlight. The process took about two weeks from start of growth to the harvesting of their so-called green crude. This green crude was then processed to yield what biofuel advocates like to call fungible fuels, those that are functionally interchangeable with 91 octane gasoline, 89 cetane diesel, and jet fuel (with a sub-47 degrees C freeze point to permit high-altitude flight).

For a while, algae appeared to be the microbe of choice for producing crude oil. The term "algae" covers a group of photosynthetic organisms that range in size from microalgae (e.g., single-cell creatures such as diatoms), to macroalgae (including large seaweeds such as kelp). Even smaller than microalgae are cyanobacteria, which are bacteria not algae.

Regardless of size, they all have the ability to turn carbon dioxide and water into carbohydrates and other products through the chemical transformations of photosynthesis (albeit across wide variations in efficiencies).

Photosynthesis is a sequence of reactions that occur in green plants and photosynthetic bacteria, in which light energy from the sun is used to produce carbohydrates and all the rest of the plant's materials. Schematically the reaction is:

carbon dioxide + water + light energy→ carbohydrates + oxygen

As a petroleum-producing organism, algae has a number of advantages. First, with the exception of its incarnation as seaweed, algae is not a food crop. Second, algae can be grown virtually anywhere that there's sunlight, and on land that's unsuitable for conventional crops—in deserts, for example. Third, it doesn't need potable water for growth, but can thrive in brackish water, seawater, or even on wastewater, meaning that it doesn't compete for the world's scare drinking water supplies. Like any other photosynthetic organism, algae consumes CO_2, meaning that it actually removes carbon dioxide, a greenhouse gas, from the atmosphere. And its end products—fatty-acids (lipids) or other oils, and even some types of long-chain hydrocarbons—can be processed into any of the three classic petroleum distillates: diesel oil, gasoline, and jet fuel.

Finally, algae can be genetically modified in an effort to maximize its efficiency or yield, or to fine-tune the chemical characteristics of its output. Given all these features, algae would appear to be an excellent biofuels production platform.

But it also has its shortcomings. To begin with, algae does not simply secrete its product in such a way that it can be siphoned off or skimmed from the top, like cream. Instead, the stuff must be separated from the organism by brute force—by centrifugation, for example. This builds in an extra layer of inefficiency, like an orange juice manufacturer that extracted the juice by hand, one orange at a time, rather than by mass-production extraction machinery.

Second, once separated, the algae-produced fatty acids must be refined, more or less like ordinary raw crude oil, into the desired fungible end products. That process takes energy, which has to come from somewhere. And if it comes from coal-burning power plants, it puts even more CO_2 into the atmosphere.

Third, algae growth requires nutrients, such as nitrogen and phosphorus, which often come from petroleum feedstocks. The microbe therefore utilizes in its growth process some of the very substances it was intended to replace.

Fourth, algae is an excellent light blocker, which means that the open ponds in which it is grown cannot be very deep. Indeed, the actual light penetration is less than 1 millimeter, which means that vigorous mixing and hundreds of gallons of water are required for every gallon of oil produced. This in turn makes for major space requirements. In fact, one study concluded that growing enough algal fuel to supply the world's entire jet fleet would require a land area the size of Maryland.

Clearly, the algae-to-biofuels road is not a smooth one, and even with the research on algal fuels now being done by Craig Venter's Synthetic Genomics (with $300 million in funding provided by ExxonMobil), it's far from certain that algae will turn out to be the preferred biofuels production platform.

ℬ   ℬ   ℬ

But as we have seen, algae is not the only microbe that can make biofuel; so can the industrial microorganism *E. coli*.

In 2005 Chris Somerville, a professor of plant and microbial biology at the University of California–Berkeley, Jay Keasling, David Berry, and I co-founded a private start-up company whose objective was to use engineered *E. coli* to produce commercial quantities of renewable diesel fuel (as well as stocks of sustainable, green chemicals). We called the company LS9 because it was the ninth life sciences firm to be funded by the venture capital group Flagship Ventures. One of the primary attractions of using *E. coli* as our production platform was that unlike algae, *E. coli* can be engineered to make its fungible petroleum products directly—the microbe does not need to be broken up in order to release its end product. Instead, we would create these fuels according to a streamlined, one-step synthesis protocol known as consolidated bioprocessing. The microbes would consume feedstock molecules and secrete the desired fuels or chemicals, which would float to the top of a fermenter column where they could be skimmed off like cream—no centrifugation, distillation, or other intermediate steps would be required. Using this protocol, making new petroleum would be as simple and straightforward as brewing beer.

The concept of genetically engineering *E. coli* to make biofuels directly is not new. In 1987, for example, a group of researchers from the University of Florida and Southern Illinois University managed to get *E. coli* to produce ethanol. They did this by taking some genes from the bacterium *Zymomonas mobilis*, which was known to produce ethanol as one of its principal fermentation products, and inserting those genes into *E. coli*. Using simple sugars as their feedstocks, the researchers found that the reprogrammed *E. coli* turned out ethanol quickly in appreciable amounts. Although sugar, a foodstuff, was consumed in the process, the group pointed out that further engineering of the microbe ought to enable it to produce ethanol using hemicellulose (inedible plant parts) as feedstock materials.

Of course, ethanol is not petroleum. But in 2010, a group of LS9 researchers published in the journal *Science* that they had found the holy grail enzymes that make alkanes (real diesel, not "biodiesel" esters) from fats. The trick was to select the appropriate genetic structures from other organisms found in nature, which the researchers did by comparing DNA from ten species of cyanobacteria that made "trace amounts" of alkanes with one species that made "undetectable amounts." To prove that these genes were correct they inserted them into *E. coli* which enabled the microbes to directly grow small research quantities of diesel oil. The next step was to scale up the process.

By this time, LS9 had a pilot plant going in South San Francisco. The heart of the operation was a 1,000-liter fermenter tank, which was soon producing larger, batch quantities of our Ultraclean Diesel, as we call it. The microbial fermentation took only three days from start to finish, and the end product, a synthetic biodiesel, was so chemically close to conventional diesel oil that it met the American Society for Testing and Materials (ASTM) standards for road use in the United States, and was found to be chemically equivalent to California clean diesel.

In January 2010, LS9 took a major step toward the mass production of diesel oil by purchasing a bankrupt biofuels production plant for pennies on the dollar in Okeechobee, Florida. The facility included four large, million-liter fermenter tanks, storage tanks, cooling tower, and a water treatment system. We aim to produce UltraClean Diesel at the rate of 50,000 to 100,000 gallons per year initially, and then to ramp up to 10 million gallons.

At first the plant will use sugar cane syrup as feedstock, but ultimately we want to use inedible hemicellulose in place of sugar, thereby enabling us to make biofuel without using any food sources. In 2010 Keasling, together with colleagues at Berkeley and LS9, published a piece in the journal *Nature* laying out the engineered metabolic pathways that would allow *E. coli* to do this. In recognition of its achievements, the Environmental Protection Agency awarded LS9 the 2010 Presidential Green Chemistry Award.

⚘   ⚘   ⚘

The current situation in biofuels is one of finding and then optimizing the major players: optimizing the microbes for high-yield, efficient production through genome engineering, matching the microbe with appropriate feedstock molecules or micronutrients, and then fine-tuning the entire production process for generating clean fuels that are drop-in ready for use in the gas tank at costs that are competitive with those of natural petroleum products. But the *E. coli* at LS9 is not photosynthetic, and anyway it's not a good idea to put all of your eggs in one basket. Consequently in 2007 David Berry (who was one of LS9's cofounders), venture capitalist Noubar Afeyan, and I formed another company, Joule Unlimited, to make fuels using what many regard as the most promising microbe family yet, cyanobacteria.

Cyanobacteria were once known as blue-green algae. They are not really algae, however, but a class of bacteria that derive their energy from photosynthesis, just as if they were plants. These photosynthetic bacteria happen to be one of the game-changing organisms in earth's history, for they are thought to have been responsible for the "great oxygenation event," a major environmental transformation that happened about 2.4 billion years ago, during the Archean. The predominant life forms at that time consisted of anaerobic organisms, which thrive in the absence of free oxygen. The metabolism of cyanobacteria, however, released free oxygen into the atmosphere, which had the dual result of wiping out most of the oxygen-intolerant organisms while simultaneously making possible the evolution of aerobic organisms (such as ourselves), which depend on free oxygen. So the fact that we exist at all is in large part attributable to the metabolic activities of cyanobacteria.

More than ten thousand varieties of cyanobacteria have been discovered. They are found in frigid Siberia and in fiery deserts; on shower curtains and in toilet tanks; in the world's oceans, and in niche environments such as hot springs, salt works, and hypersaline bays. Indeed, some biologists regard cyanobacteria as the most successful group of microorganisms on earth.

Joule Unlimited expects these organisms to be equally successful at converting sunlight into diesel fuel. During the first two years of its existence, the company operated in relative obscurity (indeed in stealth mode) out of a nondescript building on Rogers Street in Cambridge, not far from MIT. News accounts in the *Boston Globe* often referred to Joule's "secret ingredient," an unknown, heavily engineered microbe. But the company was only protecting its intellectual property until the time the founders could patent their uniquely designed cyanobacteria. This it did in 2010, with a patent titled "Hyperphotosynthetic Organisms."

These engineered cyanos, it turns out, have the ability to take sunlight, CO_2, and brackish water, and then convert these ingredients into alkanes, the molecular constituents of diesel oil. Like LS9's *E. coli* bacteria, Joule's cyanobacteria secrete their end products into the surrounding watery medium. But these microbes have the additional advantage that they require *no* feedstock molecules as raw materials: no sugar, hemicellulose, salt or pepper, just some micronutrients that act more or less like fertilizer. Otherwise, it's as if they run on sunlight alone (in a process that the company refers to as helioculture).

The result, according to company president Bill Sims, "is the world's first platform for converting sunlight and waste CO_2 directly into diesel, requiring no costly intermediates, no use of agricultural land or fresh water, and no downstream processing."

Centerpiece of the system is Joule's SolarConverter, which is essentially an inexpensive, flat, transparent solar panel through which circulate thin films of cyanobacteria suspended in a bath of water and micronutrients. CO_2 bubbles in at the bottom, and the end product—alkanes—rises to the top. Powered by sunlight, the cyanos release oxygen, sugar, and clean-burning, fossil-free diesel oil.

The process has been demonstrated in the lab, as well as in a Joule pilot plant in Leander, Texas. The company calculates that an array of its Solar-Converter panels can crank out more than 13,000 gallons of diesel fuel per acre per year. Based on an industrial-scale plant, the firm expects to be able to deliver diesel at the cost of $50 per barrel (a barrel contains 42 US gallons = 159 liters). And in line with the current fashion in the bio-fuels business of equating delivered fuels with land use, the company estimates that it could supply all of the transportation fuel requirements of the United States from a land area the size of the Texas panhandle.

ॐ   ॐ   ॐ

If anything's clear from all this, it's that we're now in a transitional period, caught between the age of fossil fuels and the age of biofuels. Fossil fuels are (probably) a product of dead microbes, organisms that took millions of years to be converted from biological entities to the hydrocarbons of crude oil. Today, by contrast, we're making basically the same thing happen, only faster, and with live microbes. These living biological microorganisms are creating those same hydrocarbons that the old dead ones did, and they're doing it directly and in a matter of days, not millions of years, and they're doing it right before our eyes, not deep down inside the ocean bedrock from which they've got to be laboriously pumped out.

Nobody knows yet which microbe, process, or company will be successful and which will fall by the wayside. The competitors are in a classic

Darwinian struggle for existence, and it's not yet obvious who the winners will be . . . or even whether biofuels are truly the wave of the future.

Still, every gasoline price hike boosts the attractiveness of biofuels. Conversely, every drop in the cost of DNA sequencing and synthesis brings us closer to being able to genomically engineer the miracle microbe that will spout cheap diesel, while at the same time scrubbing CO_2 from the atmosphere. So maybe this dream is *not* in fact too good to be true. Granted, you can't get something for nothing. But gradually, over time, the biofuels development process will become more efficient and optimized, bringing us gasoline at prices if not too cheap to meter, then at least cheap enough not to fret about every time we face the pump.

The only question then remaining will be which land mass to sacrifice: Maryland, France times two, or the Texas panhandle (or maybe all three?). Me, I vote for Texas.

CHAPTER 5

-60 Myr, Paleocene

Emergence of Mammalian Immune System.
Solving the Health Care Crisis Through
Genome Engineering

ઈઠ

On September 12, 2004, fifteen-year-old Jeanna Giese attended mass at St. Patrick Catholic Church in Fond du Lac, Wisconsin. During the service, a small bat flew down from the vaulted ceiling, struck one of the stained glass windows, and fell to the floor. Jeanna Giese loved animals. At home she had two dogs and three rabbits, and she hoped one day to become a veterinarian. She asked her mother if she could catch the bat, take it outdoors, and release it, and her mother said that she could.

Jeanna picked up the bat with her bare hands, carrying it by the wingtips, then went outside and tried to let go of it. But instead of flying off, the bat sank its teeth into the tip of Jeanna's left index finger. The bite felt like a needle prick. Jeanna then grabbed the creature with her other hand and flung it away.

Back at home, Jeanna's mother, Ann, washed the wound, which was less than a quarter inch (0.6 cm) long, with hydrogen peroxide solution. Her father, John, was a construction worker and a deer hunter and he was accustomed to all types of minor scrapes, cuts, and bruises. The bite didn't look like much, and everybody forgot about it.

A month after the incident, however, Jeanna developed the symptoms of clinical rabies: fever, blurred vision, and nervous system abnormalities. Physicians at the Children's Hospital of Wisconsin took blood, saliva, and cerebrospinal fluid samples, and sent them to the Centers for Disease Control and Prevention in Atlanta. The next day, the CDC confirmed the diagnosis of rabies.

Rabies is a disease with an almost 100 percent mortality rate for patients who are not vaccinated before symptoms appear. The human immune system is just about helpless against it, since the rabies virus kills before the body can generate a sufficient immune response. The virus replicates at very low levels, and this makes it hard for the immune system to detect its presence within the bloodstream, tissues, or cerebrospinal fluid. And so before the body can generate a stock of antibodies that would disable it, the virus has already done its work.

Which meant that if events followed their normal course in this case, Jeanna Giese would soon be dead.

ɤə   ɤə   ɤə

The mammalian immune system evolved to a high degree of maturity in the Paleocene epoch, which lasted from about 65 million to 55 million years ago. The epoch is noted for having followed the mass extinction event that marked the end of the dinosaurs (the so-called K-T boundary, for Cretaceous [Kreidezeit]-Tertiary). During this period mammalian species diversified widely and spread out across the planet. The success of the mammals was in great measure attributable to the invention of the immune system and its ability to thwart a wide variety of pathogens: viruses, bacteria, fungi, and parasites.

The immune system's ability to fight these invaders resulted from the fact that all mammals, including humans, evolved in a world full of microorganisms. This parallel coevolution of pathogen and host was a neverending battle of wits as invading organisms continuously developed new strategies for infection while the immune system invented its own new countermeasures for defeating them. The same cat and mouse game continues to this day, and left to its own devices would presumably go on into the indefinite future. Synthetic genomics, however, offers the potential to tip the balance decidedly in our favor, outstripping any future evolutionary advances on the part of the microbes.

The human immune system is an amazing collection of biological mechanisms and devices that have evolved an almost magical ability to recognize a pathogen that it has never before encountered, mount a precise and targeted defense, and rid the body of it. But with some pathogens, such as the causative agents of rabies, AIDS, herpes, malaria, and tuberculosis, among others, the immune system offers only a weak defense. Even worse, sometimes the system is oversensitive, treating essentially harmless invaders as enemies, in what is known as an allergic reaction. Further, in autoimmune diseases the body becomes allergic to its own proteins. An immune reaction can also be too aggressive. In some types of hepatitis, for example, people die not from the hepatitis virus itself, but from the immune system's destruction of the liver. Further, the immune system is responsible for organ rejection, and it also fails to protect us from most cancers. Decidedly, the human immune system, and the phenomenon of immunity itself, are mixed blessings.

It has been known since ancient times that if people contracted and recovered from certain diseases, they rarely suffered from them again. Nearly 2,500 years ago, in his *History of the Peloponnesian War*, Thucydides wrote about a plague epidemic: "No one caught the disease twice," he said, "or if he did, the second attack was never fatal."

The biological underpinnings of the immune response have been understood only recently. And not fully understood at that, because this is a system of enormous complexity, with many active agents and strategies. An account of it is not for anyone in a hurry, nor is it for those with delicate stomachs or weak nerves. First of all, when we speak of being immune to a pathogen we mean that the body has developed a mechanism for recognizing an invader as a threat, has activated a means for defeating the attacker's mode of action, and has acquired the ability to respond more strongly to that same pathogen in the future. How all of this happens is a convoluted tale suggestive of major military engagements such as the Peloponnesian War itself.

The foreign object in question is called an antigen, a biological substance (a pathogen, toxin, or enzyme) that stimulates an immune response in the body. A canonical immune response to the appearance of an antigen within the body involves three main classes of disease-fighting agents: macrophages, antibodies, and T cells of various types. An antigen floating in the lymphatic system, for example, will sooner or later enter a lymph node, where it will meet up with a macrophage (literally, a "big eater"), one of the body's natural scavengers and all-purpose trash disposal units. A macrophage gobbles up foreign, dead, deformed, or used-up body parts much as Pac-Man swallows a yellow pellet, except that when the macrophage spits out pieces of the semidigested object, the particles often adhere to its outer surface. This is a rather messy operation, with a definite resemblance to an infant's dribbling excess baby food out of its mouth.

Roving bands of T cells continuously scan the exteriors of macrophages, as if searching for trace evidence of invasive forces. T cells, which were discovered in the 1960s, are lymphocytes (white blood cells) that have matured in the thymus. There are two major types of T cells, helper T cells, also known as CD4 T cells after the CD4 molecule attached to their outer surfaces (a CD4 molecule is a type of glycoprotein), and killer T cells, the top guns of the immune system. They are also called CD8 T cells because their outsides display an identifying trademark CD8 protein molecule.

When a helper T cell catches a macrophage dribbling out antigen fragments, it will report this fact by chemical messenger (a lymphokine, such as interferon) to another type of white blood cell, a B cell, also called a B lymphocyte (B for the bursa, an organ in birds in which B cells were first discovered in the 1950s). Upon receipt of the chemical message that antigens are at large in the bloodstream and the lymphatic system, the B cell goes ballistic and essentially turns itself into an antibody factory.

If killer T cells are the body's top guns, antibodies are its stun guns. They are little Y-shaped protein devices that attach themselves to (or "bind") the antigen against which they are made. This has the effect of neutralizing it and rendering it *hors de combat*—out of action. The human body has the ability to churn out staggeringly huge numbers of antibodies, of at least 100 million different types, an attribute known as immune *diversity*. Each type of antibody, furthermore, has the special property of being able to bind to and neutralize primarily the specific antigen that triggered its formation; this is immune *specificity*. Taken together, these two features of the immune system, diversity and specificity, enable it to produce vast quantities of antibodies that are tailor-made to fight as many of the millions of different pathogens as it might one day encounter.

Antibodies are virtually the defining units of immunology. The question is, How is the human body able to produce such a great variety of antibodies without knowing in advance the nature of the specific microbial invader that it must defeat? (Among immunologists, this is known as the "generation of diversity," or the GOD problem.) To begin with, the antigenic universe is not infinite. All pathogens are made from a few different types of atoms: sulfur, carbon, hydrogen, phosphorus, oxygen, nitrogen. These are enough to produce an extremely wide variety of pathogens, but that number is nevertheless finite and limited. Antibodies are proteins, and they too are composed of a few different kinds of atoms. The problem facing the immune system, then, is how to generate a sufficient variety of antibodies to bind to and disable any and all of the different types of antigens it might meet.

Antibodies are produced in the bone marrow, where the cells that put them together generate them in random combinations, somewhat analogously to the manner in which the rotating cylinders of a slot machine click to a stop sequentially, producing a different lineup of pictures with each separate run. The business end of the Y-shaped antibody molecule (technically, its antigen-binding site) is composed of two pairs of protein chains, one of which has two parts, the other, three. The atomic structure of each part is encoded in the DNA by a separate gene segment. The chains are put together in such a way that the type of atoms composing one part can be randomly combined with those that compose the other parts so as to generate a stupendously large pool of different antibody types.

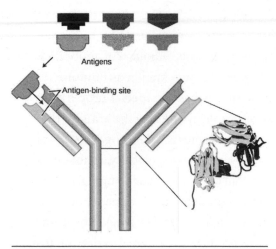

Antigens

Antigen-binding site

Figure 5.1 How antibodies consisting of four protein chains bind to antigen targets.

The success of an antibody reaction in any given immune response is a function of several factors, including the nature and virulence of the pathogen, the age and healthiness of the person infected, and even, it has been argued, that person's emotional state. In a successful operation, the bound antigens are ultimately ingested by the macrophages and cleared from the body. Viruses that have invaded cells are detected and destroyed by the CD8, or killer T cells, which send them a message telling them to self-destruct, thereby initiating the process of programmed cell death, or apoptosis (pronounced ape-oh-tosis).

The fact that billions of people are alive and well across the planet is testimony to the extraordinary robustness of the immune response over the millennia of our existence. The fact that millions die each year of infectious diseases reminds us that for all its effectiveness, the immune system is nevertheless riddled with defects. Reason enough to bring synthetic genomics into the picture.

⁂

When Congress passed a sweeping health care reform bill in 2010, everyone knew that the legislation would end up costing billions if not trillions of dollars. The specifics of the program were largely confined to customary, established, and time-honored (perhaps even hidebound) health care practices. But synthetic genomics will make it possible to go beyond standard practices in at least two ways. One is by providing entirely novel methods of treating existing diseases. Mostly, this involves genomically engineering microorganisms or nonhuman mammalian species for the purpose of performing specific medical interventions. The other is by reengineering the human genome itself for the purpose of preventing many diseases from occurring in the first place. The second application would be radical and far-reaching in its effects.

Admittedly, the notion of altering our own genes is an idea that takes some getting used to: if certain people are frightened by genetically modified foods, they're going to be stricken with existential terror at the thought of genetically reengineered human beings. But if introducing small, surgical alterations to our genome would make us immune to all viruses, known or unknown, it would revolutionize medicine. Medical treatments for viral diseases would be obsolete. The diseases just wouldn't exist (just as smallpox no longer does), and the cost savings—both in dollars and in human misery—would be immense. But synthetic genomics could radically improve human health even while leaving the human genome intact—by changing the genomes of other organisms.

One problem in treating cancer is that the chemical agents used in chemotherapy to kill cancer cells often kill healthy cells too. Scientists have proposed various schemes for delivering the chemicals selectively to the cancer tumor alone. For example, physicist Alex Zettl at the University of California–Berkeley has suggested placing the chemicals inside tiny, radio-controlled carbon nanotubes that would be guided to the tumor and then triggered to release their anticancer agents. Making this work would require a whole infrastructure of these new devices, and a nontrivial amount of engineering, but it's nevertheless possible.

But there's another way of accomplishing the same goal biologically. In 2006, J. C. Anderson and a group of molecular biologists at various research institutions in California outlined a scheme for engineering *E. coli* bacteria to attack and destroy cancer cells, thereby creating what amounts to a bacterial drug delivery system. In this case, ironically, bacteria would no longer cause disease but cure it.

Anderson and his colleagues advanced their proposal in the *Journal of Molecular Biology*, with "Environmentally Controlled Invasion of Cancer Cells by Engineered Bacteria." Here they pointed out that after being injected into the bloodstream some bacterial species, *E. coli* among them, are naturally drawn toward tumors, and furthermore are preferentially attracted toward solid tumors including bladder, brain, and breast cancers. They might do this because of disruptions in the blood flow or some other aspect of the microenvironment near the tumor cells.

To this natural partiality of *E. coli* for cancerous tumors, the researchers added two genetic features from other organisms. First they gave their microbe the ability to invade cells. They did this by taking a gene from a bacterium called *Yersinia pseudotuberculosis* that when expressed by *E. coli* would allow the latter organism to adhere to and then invade mammalian cells. They called this newly engineered, invasive organism *inv+ E. coli*, and experimentally demonstrated that it could enter skin, liver, and even bone cancer cells without difficulty.

Then, to make sure that the new organism would enter mainly cancerous calls, and not healthy cells, they programmed a second discriminating ability into *inv+ E. coli*. Cancer cells have higher densities than normal cells, and so the researchers wanted *inv+ E. coli* to be able to sense the densities of different environments. They gave it a sensor. They took from the marine bacterium *Vibrio fischeri* a genetic circuit called *lux* that enables the organism to distinguish among varying cell densities in its normal aqueous environment. Inserting the *lux* gene into *inv+ E. coli* gave the researchers a genetically rewired bacterium that could distinguish among cells of different densities and selectively invade only higher-density cells. What if the *lux* bacterial circuits mutate? The cells are only used for a brief time and a few molecular mistakes in an ocean of such noise seems acceptable. In any event, this will be extensively tested in clinical trials.

Still, that is only half the story. "Specific invasion of tumor cells is only one component of an anticancer bacterium," the experimenters said. "Once inside target cells, a cytotoxic or immuno-stimulatory response must instigate destruction of the tumor. Various bacteria have been engineered for this effect including *Salmonella* that metabolize a chemotherapeutic pro-drug at tumor sites."

Such a biodevice raises several other possibilities, because once you have a bacterium that can sense the microenvironment of a tumor, invade it, and release certain chemicals inside it, you then have a general platform for a range of similar devices that can also selectively release live vaccines, probiotics, or even genes into predefined target cells.

In the Church lab at Harvard we are working to make these invasive bacteria safer. This is important because bacterial cells can cause inflammatory responses in the host. But we can engineer the bacterial cell wall so that the cell does not elicit this response. Establishing a new counter (of cell divisions or of time) in the synthetic circuitry of the cell will ensure that these invasive cells don't over-replicate in the nutritionally rich environment of the human body. Alternatively, we can limit replication by engineering into the cell a nutritional dependence on something not present in the body (e.g., azido-phenylalanine) and by keeping this nutrient present as the bacteria grow and until they are injected into the patient. Additional safety features include engineering the genomes to be incompatible with DNA in other bacteria so that genetic information cannot flow in from or out to the environment, possibly by changing the core translational code of the cell (as described below).

The cancer-killing stealth bacterium is one scheme for improving human health through genomic engineering. A second is to custom-build antibodies to fight specific diseases. Such antibodies can then be produced in mass quantities and introduced into the bloodstream of patients suffering from those conditions.

One of the most promising advances in medical biotechnology was the development during the 1970s of monoclonal antibodies. Monoclonal antibodies (mAbs) are identical to each other because they are made by immune system cells that are clones of a unique parent cell. Because of their extreme specificity of effect, monoclonal antibodies were thought to be "magic bullets" against various illnesses.

The original idea was to produce mAbs in vitro in a somewhat roundabout process starting from mice that had been injected with a specific antigen, and ending with vials of identical antibodies that would be tailored to neutralizing that antigen. The antibodies would then be injected into human patients, immunizing them to the disease in question.

There was a problem, however—the monoclonal antibodies that had originated within mice were rodent antibodies, not human antibodies. For that reason, they were rejected by the patient's immune system and rapidly eliminated from the bloodstream. The patient's immune system produced human antimouse antibodies against the mouse-derived monoclonal antibodies, since it regarded them as antigens. The magic bullet, in other words, backfired and failed to clear the body of the disease (but at least did not kill the patient).

But there was a second act to the drama. Scientists had been producing so-called chimeras, or transgenic species, in which the genes from one organism were merged with those of another, since the 1970s. The first transgenic species were bacteria, but biologists soon progressed to inserting foreign genes into the genomes of animals such as mice, sheep, or cattle, for various research, medical, or commercial (or even novelty) purposes. Later, in the early 1990s, other researchers, including Nils Lonberg at the California medical biotech firm Medarex, realized that it would be possible to insert enough human gene sequences into a laboratory mouse genome to create a transgenic, or "humanized," mouse. (Lonberg and I were grad students together in Walter Gilbert's lab at Harvard.) The theory was that monoclonal antibodies derived from a humanized mouse would not be recognized as foreign by a human recipient's immune system, and therefore would not be rejected by the body.

As it turned out, the second magic bullet scheme worked, and by 2005 more than thirty-three humanized mouse-derived mAb drugs were in various stages of clinical trials, including several developed by Lonberg, whose company, Medarex, had trademarked the name "HuMAb-Mouse" for its proprietary strain of humanized mice. Inasmuch as its mouse genes for creating antibodies had been inactivated and replaced by human antibody genes, the HuMAb-Mouse had acquired the ability to make antibodies that were fully human.

For Nils Lonberg, the payoff came in 2009 when the Food and Drug Administration approved four HuMAb-Mouse drugs for use in humans, for various illnesses including rheumatoid arthritis, autoinflammatory syndrome diseases, psoriasis, and chronic lymphocytic leukemia.

"The four drugs approved in 2009 really represent only the tip of the iceberg for our transgenic mouse platform," Lonberg said in 2010. "There are a lot of exciting drugs behind these in clinical development, and we continue to use the platform for drug discovery."

෯෨   ෯෨   ෯෨

Cancer-seeking stealth bacteria and humanized mammals for monoclonal antibody production show that genomic engineering can improve human health. Still, these two technologies are mopping-up operations, interventions employed only after a disease occurs. A better application of synthetic genomics would be to prevent diseases from ever occurring by altering the human genome, directly reengineering ourselves to higher levels of health, hardiness, and disease resistance.

If that sounds heretical or insane, consider that human beings have been doing something like this for a long time, even with no knowledge of genetics. When people try to select a mate possessing the best possible qualities—strength, health, stamina, good looks—for the purpose of producing healthy offspring, the effect, if and when achieved, would result in the creation of a new genome, one whose genes yielded the sought-after physical traits. In more recent history, when women seeking to give birth to "genius" babies applied to the now defunct Repository for Germinal Choice (popularly known as the Nobel Prize sperm bank) for insemination with what they thought was genius-level sperm cells, what they were trying to do was to incorporate high intelligence genes into their children's genomes. (Whether this actually works is another matter.)

So it's not as if we don't already try to selectively alter, improve, and beef-up human genomes. Synthetic genomics attempts to do the same thing systematically while drawing on a detailed scientific understanding of genetics, as well as huge experimental and informational databases pertaining to any proposed genomic alteration.

One of the greatest human health innovations of all time would be to make ourselves multivirus resistant—render ourselves immune to all viruses, known or unknown, whether currently existing or waiting in the wings to come off the evolutionary assembly line in the future. This means that there would be no more influenza pandemics, no more common colds, no more HIV-AIDS, no more herpes, hepatitis, or polio. And absolutely no more rabies.

We have already seen how we could make human beings multivirus-resistant by mirror-flipping every chiral molecule in our bodies. We could also achieve multivirus resistance by changing our genetic code.

And here we are, at the climax to "the greatest story ever," the story of the genome. For the genome is written in code, a code that was established by nature and has remained unchanged for billions of years. While countless individual genomes have come and gone during that time, the code on which all of them were, and still are, based has remained unchanged for eons. But now, after all those billions of years, we propose . . . to change it.

Yes, this sounds like hubris all right, not to mention "playing God." It's almost as if we're proposing to reposition the stars in the heavens, or to rewrite the periodic table of the elements.

In reality, there's nothing blasphemous about the prospect of changing the genetic code to improve human health and prevent illness and death from viral infection. The process is probably easier than changing ourselves into mirror people. There is only one (hard) way to change ourselves into mirror humans, but there are a vast number of ways of changing our genetic code.

We've seen that viruses are able to replicate themselves inside our cells because they use the same genetic code, or speak the same genetic language, as our own cells. One way to defeat the virus's ability to replicate, then, would be to throw a monkey wrench into this chain of events by altering the genetic code of the host cell, as well as that of the cellular machinery that translates the virus's codons, so that the virus's genome becomes unreadable by the host cell and its translation apparatus. More specifically, we're going to alter some of the codons of the host cell's genome, as well as some codons within the cell's translation apparatus, so that these cellular mechanisms no longer recognize the codons of the virus's genome. Instead, the host cell's translation mechanisms will trip up on the virus's codons and effectively come to a halt. At that point, the virus is stopped in its tracks. The viral genetic message that formerly made perfect molecular sense to the host cell now amounts to nonsense, and as far as the host cell is concerned, the viral invader's genome is incomprehensible molecular babble. The upshot: multivirus resistance—immunity to all viruses.

Making changes in our genetic code is not a matter to be taken lightly. The process would start with experiments on bacteria and proceed to ever more complex genomes to mammals before ever trying the procedure on human beings. So we will proceed cautiously, and with due regard to safety. We are working this out initially for industrial microbes like *E. coli*, which are especially susceptible to contamination with viruses (phages), a susceptibility that endangers production vats of millions of liters of pharmaceuticals or chemicals. Such a contamination event occurred with hamster cells at Genzyme in Boston in 2009—and it probably happens more often than we know, since companies don't like to publicize their mistakes.

So, how do we change the genetic code? By removing a key part of the host machinery that the virus needs to replicate itself. But doing that and nothing else, however, would also make the *cell* unable to replicate itself. You can't remove a key part of the host machinery that the virus needs to replicate itself until you have removed every aspect of the cell's genome whose replication also depends on that part. In the following section I'll illustrate what this means with an example that has nothing to do with the genetic code but nevertheless exemplifies the principle stated above.

The Parable of the Locks

Once upon a time a pawn shop chain called Buy 'n' Cell installed indestructible metal roll-down night gates and equipped them with heavy-duty combination locks for each store. To keep things simple for the employees who opened the shops in the morning, management left the locks at the factory setting of 123456789. But a local gang, aptly called the Striking Virals, wanted to break into these shops to enrich themselves and maybe become even bigger and better and more powerful than before, and go on to raid even more shops.

One night, gang members went to each store and tried the combinations 000000000 and then 123456789. The second number worked perfectly, and the gang looted each and every shop. The next morning management changed all of the locks to 223456789 but didn't manage to reach all of the employees with the new number. Those stores didn't open. This illustrates the all-important metaphysical principle that you can't change the combination that the employees need to open the store unless you have informed every store employee who depends on knowing that combination in order to open the shop.

The next night, the Virals tried the old combination without success but soon came up with the new combination, since it was only one mutation away from the original. The Buy 'n' Cell staff responded the next day by changing the combination to 892220611 and giving the new number to all employees. Later that night, after a few minutes of trying different numbers, a genius-level member of the gang realized that even going at the rate of one try per second, nonstop, it would take them half a billion seconds (fifteen years) to try even half of all the possible combinations. So they gave up and retired from the scene, but not before covering all the metal gates with graffiti copies of Napoleon's famous epigram, "Glory is fleeting but obscurity is forever."

So how does this stirring tale relate to the process of changing the genetic code?

We start with transfer RNA (tRNA)—folded molecules that transport amino acids from the cytoplasm of a cell to a ribosome that strings together amino acids into proteins. Each tRNA molecule is essential for the life of the host, unless you can get rid of every instance of the codon that needs that particular tRNA to transport an amino acid to the ribosome. To get rid of those codons you'd need to swap each of them with a so-called synonymous codon that uses a different tRNA but translates to the same amino acid. Or, in terms of the parable, to do that you'd need to swap that combination (codon) with a so-called synonymous combination (codon) that uses different numbers (codon letters) but performs the same function (prompting a tRNA to transport the same amino acid to the ribosome). (A synonymous codon is one whose different nucleotides encode the same amino acid; for example, both AGA and CGG encode the amino acid arginine.)

Our first goal in the lab was to remove the gene for the protein called "release factor 1" (nicknamed RF1). This protein is required for *E. coli* ribosomes to stop adding amino acids at the end of making any of 322 different proteins that terminate in the stop codon UAG. RF1 binds to the mRNA at the UAG and causes the ribosome to stop. We couldn't delete RF1 until we changed all 322 instances of UAG to the synonymous UAA codon (which is handled by a different protein known as RF2). When an invading virus comes along, it needs the RF1 to use the UAG codon, but RF1 isn't there any-

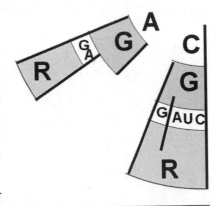

Figure 5.2 A small segment of the genetic code (from Figure 3.2) showing all the six synonymous codons for the amino acid arginine (R): AGA, AGG, CGC, CGU, CGA, and CGG.

more! The ribosome will still manage to function, but badly. Nevertheless, probably two or more codon changes will be needed to defeat all viruses.

When we began these experiments we weren't sure that these so-called synonymous codons would all be truly equivalent, because living things tend to use their parts in many gloriously complex and overlapping ways. Ways that would be considered "spaghetti code" if a programmer ever tried such undisciplined and tortuous protocols. However, we did successfully make all of the 322 changes. For our synthetic purposes (not necessarily for all of the natural purposes of the cell), these changes are innocuous and hence we can take the next steps toward multivirus resistance.

How did we manage to make so many changes? The answer lies in the process and instrument called MAGE, introduced in Chapter 3. That process allows us to make many changes simultaneously and in parallel.

Next we target the tRNA that recognizes the AGA and AGG codons, normally translating these to the amino acid arginine (R). In Figure 5.2 we see the six synonymous codons for arginine. If every one of the AGA and AGG codons is changed to CGA, CGG, CGC, or CGU, all of which *also* code for arginine, then we can remove the gene for the tRNA that

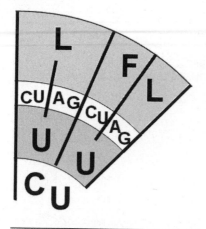

Figure 5.3 A small segment of the genetic code (from Figure 3.2) showing all six synonymous codons for the amino acid leucine (L): CUC, CUU, UUC, UUU, UUA, and UUG.

binds, recognizes, and translates the AGA and AGG codons (otherwise removing them would be lethal to the cell). If that tRNA is missing, the ribosome will stall at each AGA and AGG and make defective versions of essential cell proteins. Even if other tRNAs jump in and prevent stalling, the number of incorrect amino acids would be disastrous to the cell. Such genome-wide reassignment of codons can be extended to still other codons and then the corresponding tRNAs can be deleted. For example, leucine (L), like arginine, has six synonymous codons.

With regard to the piece of genetic code in Figure 5.3, we can now free up (make nonessential) one of the three tRNA molecules for leucine by moving all of the CUC and CUU codons to CUA or CUG. At that point, the host has freed up (eliminated all uses of) five of the sixty-four codons (UAG, AGA, AGG, CUC, CUU) and we have deleted three genes that are normally essential to translation and hence are essential for life. This means that invading viruses not only can't use the cell machinery but they don't even have a chance of mutating to use it.

As a concrete, real-life example of this, one of the smallest functional peptides in all of biology has the five amino-acid sequence MRLFV (methionine-arginine-leucine-phenylalaline-valine, which provides resistance to the antibiotic erythromycin). If the peptide were initially encoded by the following five triplets,

AUG *AGGCUU*UUUGUG*UAG*

then after recoding, the sequence would be:

AUG *CGGCUA*UUUGUG*UAA*

(altered codons in black boxes), still making the same old MRLFV peptide, because the changed codons are synonymous with and perform the same function as the original ones. But if a tiny gene like the original one above were required for an incoming virus to be replicated, the virus would still be using the original five codons; and if the host had repurposed the original three, then the virus would find (much to its surprise!) that three of its precious codons (**AGG,CUU,UAG**) no longer worked.

We are in the midst of changing a dozen codons genome-wide in *E. coli* (in parentheses is the one-letter abbreviation for the amino acid or stop encoded, together with the number of times each is used in the small viral genome below): ACC (T, 24), AGA (R, 10), AGG (R, 6), AUA (I, 19), CCC (P, 10), CGG (R, 12), CUC (L, 27), CUU (L, 19), GCC (A, 20), GUC (V, 21), UAG (stop, 2), UCC (S, 20). This recoding strategy would affect 190 codons of the 1,147 total in the virus below. Most viruses use those twelve codons even more times in essential functions and "expect" a good host (or hostess) to provide the tRNA molecules that recognize them. But after our recoding, the essential viral RNA molecules stall on the ribosomes as soon as any of the missing tRNA molecules are needed, and the game is over.

Figure 5.4 is the full sequence of one of the smallest known viruses (MS2)—the first genome ever sequenced (in 1976 by Walter Fiers and team). It's related to the first genome subjected to Darwinian molecular evolution in the lab (Sol Spiegleman's monster Q-beta in 1965). The natural MS2 genome encodes four known proteins, which (in order along the genome, which is also essentially the mRNA) are: the *maturation protein* (one copy per virus) that holds onto a unique point in the RNA genome and guides it into the shell coating the virus and later into the host cell; the *major coat protein* (180 copies form the outer shell of the virus); a *lysis gene* used to help the baby viruses get out of the trashed cell host; and finally, the *gene for the polymerase* that makes copies of the genome. Interestingly, the lysis gene extensively overlaps the coat and replication proteins (underlined) so those base pairs are duplicated in parentheses in Figure 5.4. The protein coding regions are indicated by spaces between the triplet codons. The regions without spaces are not translated into proteins but are useful in regulation of timing and levels of the proteins.

gggugggaccccuuucgggguccugcucaacuuccugucgagcuaaugccauuuuuaaugucuuuagcgagacgcuac
cauggcuaucgcuguagguagccggaauuccauuccuaggagguuugaccu

Maturation protein: gug cga gcu uuu agu ACC CUU gau AGG gag aac gag ACC uuc
GUC CCC UCC guu cgc guu uac gcg gac ggu gag gcu aac uca uac ucu uua
aaa uau cgu ucg aac ugg acu CCC ggu cgu uuu aac ucg acu ggg GCC aaa acg aaa
cag ugg cac uac CCC ucu ccg uau uca CGG ggg gcg uua agu GUC aca ucg AUA gau
caa ggu GCC uac aag cga agu ggg uca ucg ugg ggu cgc ccg uac gag gag aaa GCC
ggu ugg gcu uuc UCC gac gca cgc UCC ugc uac agc CUC uuc ccu gua agc caa
aac uug acu uac auc gaa gug ccg cag aac guu gcg aac ACC gcg ucg ACC gaa GUC
cug caa aag GUC ACC cag ggu aau uuu aac CUU ggu guu gcu uua gca gga GCC AGG
ucg aca GCC uca caa CUC acg acg caa ACC auu gcg CUC gug aag gcg uac acu GCC
gcu cgu cgc ggu aau ugg cgc cag gcg CUC GCC cua aac gaa gau cga
aag uuu cga uca aaa cac gug GCC ggc AGG ugg uug gag uug cag uuc ggu ugg uua
cca cua aug agu gau auc cag cgu uca uau gag aug CUU acg agg guu cac CUU caa
gag uuu CUU ccu aug AGA GCC gua cgu cag GUC ggu acu aac auc aag uua gau auc
cgu cug uau cca gcu gca aac uuc cag aca acg ugc aac AUA ucg cga cgu auc
gug AUA ugg uuu uac AUA aac gau gca cgu uug gca ugu ucu cua ggu auc
UUG aac cca cua ggu AUA gug ugg gaa aag gug ccu uuc uca uuc guu GUC gac ugg
CUC cua ccu gua aac aug CUC gag ggc CUU acg GCC CCC gug gga ugc UCC uac
aug uca gga aca guu acu gac gua AUA acg acg gag GUC auc AUA acg gcu gac gcu
CCC uac ggg ugg acu gug gag AGA cag ggc acu gcu aag GCC caa auc uca GCC aug
cau cga ggg gua caa UCC gua ugg cca aca acu ggc gcg uac gua aag ucu ccu uuc
ucg aug GUC cau ACC uua gau gcg uua gca uua auc AGG caa CGG CUC ucu AGA UAG

agcCCCucaaccggaguuugaagc

Coat protein: aug gcu ucu aac uuu acu cag uuc guu CUC GUC gac aau ggc gga acu
ggc gac gug acu GUC GCC cca agc aac uuc gcu aac ggg GUC gcu gaa ugg auc agc
ucu aac ucg cgu uca gcu gcu uac aaa gua ACC ugu agc guu cgu cag agc ucu gcg
cag aau cgc aaa uac ACC auc aaa GUC gag gug ccu aaa gug gca ACC cag acu guu
ggu ggu gua gag CUU ccu gua GCC gca ugg cgu ucg uac uua aau aug gaa cua ACC
auu cca auu uuc gcu acg aau UCC gac ugc gag CUU auu guu aag gca aug caa ggu
CUC cua aaa _gau gga aac ccg auu CCC_ uca gca auc gca gca aac UCC ggc auc uac
uaa

uagacgccggccauucaaacaugaggauuaccc

(Lysis protein: **aug** gaa ACC cga uuc ccu cag caa ucg cag caa acu ccg gca ucu
acu aau AGA cgc CGG cca uuc aaa cau gag gau uac cca ugu cga AGA caa caa AGA
agu uca acu CUU uau gua uug auc uuc CUC gcg auc uuu CUC ucg aaa uuu ACC aau
caa uug CUU cug ucg cua cug gaa gcg gug auc cgc aca gug acg acu uua cag caa
uug CUU acu uaa)

RNA polymerase: _aug ucg aag aca aca aag aag uuc aac ucu uua ugu auu gau_ CUU
ccu cgc gau CUU _ucu_ CUC _gaa auu uac caa uca auu gcu uac_ GUC _gcu acu gga agc_
ggu gau ccg cac agu gac gac uuu aca gca auu gcu uac uua AGG _gac gaa uug_ CUC
aca aag cau ccg ACC uua ggu ucu ggu aau gac gag gcg ACC cgu cgu ACC uua gcu
auc gcu aag cua CGG gag gcu gau ggu aau gac cgc ggu cag AUA AUA gaa ggu uuc
uua cau gac aaa UCC uug uca ugg gau ccg gau guu uua caa ACC agc auc cgu agc
CUU auu ggc aac CUC CUC ucu ugc uac cga ucg ucg uug uuu ggg caa ugc acg uuc
UCC aac ggu gcu ccu aug ggg cac aag uug cag gau ggg ccu uac aag aag cuu
gcu gaa caa gca ACC guu ACC CCC cgc gcu cug AGA gcg gcu cua uug GUC cga gac
caa ugu gcg ccg ggu ugg uGA cac gcg GUC cgc uau aac gau uca uau gaa uuu AGG
CUC guu gua ggg aac gga gug uuu aca guu ccg aag aau aau aaa AUA gau CGG gcu
GCC uag cgc CGG CUC aaa UCC guu guu gau AUA gac cug gau aca acc cag cgu
cug gcu cag cag ggc agc gua gau ggu ucg CUU gcg acg AUA gac uua ucg ucu gca
UCC gau UCC auc UCC gau cgc cug gug ugg agu uuu CUC cca cca gag cua uau uca
gaa cua uuu UCC aca aug gga aau ggg uuc aca uuu gag cua gag UCC aug AUA uuc
ugg gca AUA GUC aaa gcg ACC caa auc cau uuu gac aac GCC gua ACC AUU gcu ggc auc
uac uac ggg gac gau auu AUA ugu CCC agu gag auu gca CCC cgu gug cua gag gca CUU
gag agc ugc ggc gcg cac uuu uac cgu gGU gua GUC aaa ccg gug uau uac cuc aag
aaa ccu guu gac aau CUC uuc GCC cug aug cug AUA uua aau CGG cua CGG ggu ugg
gga guu GUC gga gug uca gau cca cgc CUC uau aag gug ugg gua CGG CUC UCC
UCC cag gug ccu ucg aug uuc uuc ggu ugg acg gac GUC gcc GCC gac uac uac gua
gcg gau ACC gau ACC cuu ggu uuc cgu CUU gcu cgu auc auc ccu cga gaa cgc aag uuc
uuc agc gaa aag cac gac agu ggu cgc uac AUA gcc ugg uuc aac uca acg agu gaa
auc ACC gac agc aug aag UCC GCC ggc gug cgc guu AUA cgc acu ucg gag ugg cua
acg ccg guu CCC aca uuc ccu cag gag ugu ggg cca gcg agc ucu ccu CGG UAG

cugaccgagggaccccguaaacggggugggugugcucgaaagagcacggggugcgaaagcgguccggcuccacgaaa
ggugggcggcuucggcccagggac

cuccccuaaagagaggacccgggauucucccgauuugguaacuagcugcuuggcuaguuaccaccca

In Figure 5.4, the twelve codons that we are changing in the host (*E. coli*) of this virus are in uppercase and black-boxed. You can see how unlikely it is that this virus would get exactly these 190 changes all at once.

An essential condition here is that you must change the *E. coli* cell's genome while the virus isn't present in the cell. If you do it slowly while the virus is inside the cell, as in natural evolution, then the virus keeps pace step by step with every bacterial innovation in a kind of arms race. By doing this recoding "offline" we can introduce something that requires so many changes all at once in the virus that it can never get enough random mutations made simultaneously to accommodate the new state of the cell. Indeed, the changed cells should now be resistant to viruses that they have never seen before. In the best case for the virus, it would have a small genome (fewer chances to get one of the missing codons), a high mutation rate (to get rid of the newly impotent codons), and a large viral population (more ways to win). The smallest viruses that don't depend on other viruses are about 3,000 base pairs in length. Let's say (very generously) that would be 10 places in the 3,000 base pairs that need to change to deal with the new host code. I explained how high a mutation rate can go in Chapter 3, so let's pick an aggressively high rate of one error in 1,000 base pairs or about three errors per viral genome. That means 5 percent will have no mutations and 0.03 percent will have ten mutations or more. Of those with ten mutations, only one in 10,000 will be viable and of those, one in 10^{24} will have the correct ten changes in the genome. So a population of 10^{28} of that species of virus would be needed to find one that could handle the new host. But that is on the order of all virus particles of all types in the world today. The odds get worse fast for the virus if more than ten changes are needed, and exponentially worse with the number of required changes.

E. coli has about 4,200 protein coding genes. Once we've been successful with *E. coli* and other industrial microbes, then we can tackle agricultural plants and animals that have some 20,000 protein genes—only five times that of *E. coli*. The changes would be introduced to the germ lines of the animals so that all of their progeny would also be resistant to all viruses. This is typically done by engineering pluripotent stem cells in culture and then implanting them into blastocyst stage embryos (one in the later stages of cleavage).

If all of the above practical applications work out well, then the temptation to apply the same technique to humans, and to make ourselves multivirus resistant, will be quite strong. Initially this might be done on stem cells for adult therapies; for example, blood stem cells are routinely transplanted to cure blood diseases like cancer. If you could offer stem cells that are both cancer free and virus resistant, they might be more desirable than the same cells that are merely cancer free. Indeed, there will be a demand for those cells to be not just cancer free but cancer resistant too. That would require another set of genome changes almost as complex as those needed to produce viral resistance.

Possible unintended negative consequences of using this technique would also have to be taken into account; for example, nutritionally enslaving the changed cells so that they can't escape from the lab and beat all natural relatives (which are still virus-limited). But the potential payoff of multivirus resistance is so huge that this is a procedure that almost demands to be tried.

ʃ ʃ ʃ

Multivirus resistance, if and when it's engineered into the human genome, does not necessarily make us immune to nonviral diseases such as those caused by parasites (e.g., malaria), bacteria (e.g., tuberculosis), or prions (e.g., mad cow disease). The pathogens that cause these diseases don't work by replicating themselves using human translation machinery of the human host cell. A speculative approach that could, in principle, provide protection from both viruses and cellular parasites is the mirror cell idea mentioned in Chapters 1 and 3. This would be much more challenging than making a simple mirror bacterial cell or making a viral-resistant human using code changes, but it's probably feasible. All of the tricks that pathogens use to evade the human immune system would fail in mirror humans since those tricks depend on specific chiral receptors. Furthermore, the enzymes they need to poke holes and digest the polymers of the host (DNAses, RNAses, proteases, lipases, etc.) employ high chiral specificity and hence would be useless, and nearly all of the favorite foods needed to sustain those pathogens would be unavailable.

Finally, supposing that one day we have managed to conquer an impressive array of human maladies, including infectious disease of almost every kind, we might then turn our attention to a future era in which members of the human species might occupy an entirely new ecological niche . . . on another planet. Mars, for example.

One of the primary hazards of space travel is the threat of radiation damage. Even on earth, ultraviolet radiation causes skin cancers ranging from the relatively benign basal cell carcinoma to the relatively deadly malignant melanoma. And that's with the shielding effects of the earth's atmosphere already in place and working for us. Departing from that semiprotective buffer zone is to enter a realm that's especially hostile to human health.

Radiation in open space is of two types: electromagnetic radiation, which consists of massless photons, and high-speed particles that have mass, including atomic nuclei and electrons. Electromagnetic radiation travels at the speed of light, and the various types are distinguished by their wavelengths (the distance from one wave crest to the next) and frequencies (expressed in kilohertz, or kHz). The shorter the wavelength, the higher the frequency, and the greater the energy carried by the photon. X rays and gamma rays are the highest frequency forms of electromagnetic radiation and can do the greatest damage to human cells, tissues, and DNA. Gamma rays can penetrate the body and rip apart DNA molecules and damage sperm and egg cells as well as the reproductive organs. Gamma rays are used in medicine to kill cancer cells.

Gamma rays are a form of ionizing radiation—they carry enough energy to strip electrons from the atoms or molecules to which they are attached. They affect the body's chemistry in a way that can lead to serious health problems, including cancer, tumors, and DNA damage.

There's a lot of radiation in space, and people traveling to Mars would experience six months' worth of radiation every twenty-four hours. This means that a one-year outbound voyage would subject the traveler to a dose of radiation equivalent to 180 years on the home planet. Clearly spacefarers are going to need radiation protection.

Lead shielding is one way to do it. Unfortunately, the amount of lead required might make the spacecraft too heavy to reach escape velocity. Fortunately, there are radiation-resistant animals. If we could find the genes responsible for their ability to resist radiation, then perhaps we could import those genes into our own genome and become radiation-resistant ourselves.

The all-time champion radiation-resistant organism is the extremely hardy bacterium *Deinococcus radiodurans*, a creature whose very name means "radiation-enduring." It is listed in the *Guinness Book of World Records* as the world's toughest bacterium, and has been informally nicknamed "Conan the Bacterium." It was discovered in 1956 by researchers conducting food sterilization experiments using high-dose gamma radiation, which it survived easily. This microbe can withstand a thousand times the gamma radiation that would kill a human being. Although its genome has been completely sequenced, bacteria are phylogenetically too distant from the human lineage for their genes to be easily usable.

But there is a backup radiation-resistant organism that happens to be an animal, the Bdelloid rotifer. This is a class of small invertebrates that live in freshwater pools, on the surfaces of mosses and lichens, and even in habitats that periodically dry up. To survive extended periods of desiccation, these animals have evolved the ability to repair portions of their DNA that gets damaged under these harsh conditions. In 2007, my Harvard colleagues Matthew Meselson and Eugene Gladyshev performed a series of experiments in which they subjected two species of these rotifers to high levels of ionizing radiation. They survived these megadoses well enough to repair their own double-strand DNA breaks and to produce viable offspring. (A human being would be killed by a hundredth of the radiation levels tolerated by the rotifers.) Note also the humble midge fly *Polypedilum vanderplanki* mentioned in Chapter 3 and Epilogue.

If we could locate the genetic sequences that code for the rotifer's extraordinary DNA self-repair abilities, it would theoretically be possible to isolate and then import them into our own genome, with the result that we too would have those same genomic talents and consequently increased resistance to ionizing radiation.

৯৯   ৯৯   ৯৯

That trick may help get us to other planets. Meanwhile, back here on earth we continue to suffer from thousands of diseases, including rabies.

In 2004, when Jeanna Giese entered the Children's Hospital of Wisconsin, a team of eight physicians headed by Rodney E. Willoughby consulted with Jeanna's parents about possible courses of treatment. They proposed an aggressive approach based on an untested strategy that might ultimately fail. Even if the patient were to survive, they said, she might wind up severely disabled. With essentially no other alternatives, Jeanna's parents told them to go ahead.

They planned to induce a coma to put Jeanna's central nervous system on hold and protect the brain from injury while the patient's body mounted a native immune response to the virus. In addition they would administer one of the few antiviral drugs in existence, ribavirin, along with a cocktail of sedatives to maintain the coma and ensure suppression of the nervous system.

A week into the treatment, a lumbar puncture revealed an increase of rabies-specific antibody in both the bloodstream and the cerebrospinal fluid. Gradually the patient was brought out of the coma. After about two weeks of treatment, Jeanna blinked, opened her eyes, and moved them. Two days later she raised her eyebrows in response to speech. "On the 19th day," according to the report published in the *New England Journal of Medicine*, "she wiggled her toes and squeezed hands in response to commands, [and] fixed her gaze preferentially on her mother." Some nervous system abnormalities developed as she was brought back to full conscious awareness, but she continued to progress.

"Given her continued neutralizing antibody response to rabies virus in the cerebrospinal fluid and blood, and our inability to isolate the virus or detect viral nucleic acid in saliva, the patient was considered cleared of transmissible rabies and removed from isolation on the 31st day." A month later, she was released from the hospital.*

Jeanna eventually returned to school and had no difficulties with learning or memory, although she still suffered some weakness in her left hand and foot, and walked with a lurching gait. Today, eight years after being bitten by a rabid bat, Jeanna Giese appears in more than a dozen videos on YouTube, in which she looks and sounds healthy and pleasant. She even has her own website: jeannagiese.com.

Her experience is a case study in the powers and limitations of the human immune system. She was one of only two survivors out of twenty-five attempts at using the first Milwaukee (or Wisconsin) protocol, as Willoughby's pioneering procedure is now known, and two out of ten in the revised version. Plainly there is ample room for improvement. If synthetic genomics were used to enhance our immune response, we would possess a deliberately engineered superimmunity to a vast array of diseases. This would represent a fundamental advance over the immune systems that we were natively endowed with, and which were born during the Paleocene.

Today's cutting-edge practice as applied to Jeanna Giese might one day seem incremental and brutal as we look back on the remarkable alternatives described in this chapter. All the more reason to embrace them.

---

* In June 2011, an eight-year-old California girl became the third person in the United States known to have recovered from rabies infection through the use of the Wisconsin protocol by a team of physicians, nurses, and therapists at UC Davis Children's Hospital.

CHAPTER 6

-30,000 YR, PLEISTOCENE PARK
Engineering Extinct Genomes

ℬℬ

When people think about extinct species such as the dodo, the passenger pigeon, or Tyrannosaurus Rex, what often comes to mind is the old adage, "Extinction is forever." Only it isn't. As we have seen, one member of an extinct species has already been brought back to life, albeit briefly, and it wasn't even through genomic engineering.

The Pyrenean ibex is a type of mountain goat also known as the bucardo. When they were still in existence, bucardos were one of Europe's most striking wild animals, with handsome faces, distinctive curving horns, and short, thick wool. They roamed all over the Pyrenees mountain range along the border between Spain and France. But during the late nineteenth century they were massively hunted, and by 1900 fewer than one hundred of the animals were left. The Spanish Ministry of Environment confined a small remaining population of forty bucardos to Ordesa National Park, a spectacular mountainous region in Huesca province.

In 1993 only ten individuals were left, and by 1999 there was only one, a twelve-year-old female named Celia. In the spring of that year, two biologists at the Center for Agro-Nutrition Research and Technology in Aragon, Jose Folch and Alberto Fernández-Árias, captured Celia, took a tissue scraping from her ear for the purpose of preserving the bucardo cell line, and put a radio-tracking collar around her neck. In the laboratory, the researchers multiplied the cells and then stored them away for safe-keeping in liquid nitrogen. Less than a year later, on January 6, 2000, Celia died.

According to the conventional wisdom, that should have been the end of the matter; one more species gone forever. But Folch and Fernández-Árias had a plan for bringing the animal back through a process known as nuclear transfer cloning. Nuclear transfer cloning was the same technology that gave us the first cloned mammal, Dolly the sheep, in 1996. In an ironic twist of fate, Dolly had not been cloned from the cells of a living sheep. Rather, she had been produced from a frozen udder cell of a six-year-old ewe that had died three years prior to Dolly's birth. Dolly had been literally raised from the dead. But if a live sheep can be cloned from a dead one, then why not a mountain goat? It made no difference to the frozen cells that they happened to be the last of the line: cells were cells, and so long as they contained intact DNA and the other normal structures within the cell nucleus, then they ought to be acceptable candidates for cloning.

Nuclear transfer cloning was still in its infancy, with many more failures than successes. Conceptually, the process was simple enough: take the nucleus from a cell of the animal to be cloned, transfer it to an embryonic cell from which the nucleus had been removed (an "enucleated" cell), and then implant that newly re-nucleated embryonic cell into the uterus of a surrogate mother. In theory, supposing that neither the recipient cell's cytoplasm or other organelles, nor the transferred nucleus itself, were damaged in the process, the procedure ought to work. In practice, it mostly didn't in the first few experimental trials. Apparently the process of disrupting cells in this gross manner often injured them beyond repair.

The first animal to be cloned by nuclear transfer was a northern leopard frog, produced at a research institute in Philadelphia in 1951 by experimenters Robert Briggs and Thomas J. King. They moved the relatively large DNA-bearing nuclei into the large frog eggs with a simple hollow glass pipette. During their first few attempts, the implanted embryonic cells withered and died. But the researchers persisted, and over the course of 197 attempts, Briggs and King were able to turn out twenty-seven tiny tadpole clones.

Forty-five years later, anyone might think that the nuclear transplantation batting averages would have improved. In fact, the reverse was true. When Ian Wilmut tried to clone the first sheep from an adult mammalian cell, he started out with an initial pool of 277 mammary gland (udder) cells taken from a six-year-old Finn Dorset ewe. Those 277 cells yielded up only twenty-nine embryos. Those twenty-nine embryos were transferred into the uteruses of Scottish blackface ewes—the recipient surrogate mothers. This resulted in only thirteen pregnancies. And out of the thirteen, only one cloned little lamb was born, Dolly.

So when Jose Folch, Alberto Fernández-Árias, and an international team of experts tried to clone the extinct Pyrenean ibex, they had no illusions about their prospects for success. Indeed, they were well prepared for the series of failures that they experienced. For one thing, Celia had been the last of the species, which meant that the surrogate mothers would be of a different species than Celia. That constituted no insuperable barrier, however, because interspecies nuclear transfer cloning was already a going concern. In 2001, for example, a common domestic cow was successfully used as a surrogate mother for the cloning of a wild ox. Domestic dogs have been used as surrogate mothers for gray wolf clones (although it took 372 embryos to get three live wolves), and so on.

On a date that has never been reported in the scientific literature or in news accounts, and that we here disclose for the first time, the experimenters began their first species regeneration attempts, their so-called Experiment One, in the fall of 2002.* First they had to turn Celia's somatic (skin) cells into embryos, a delicate biological black-arts process. To do it, they needed a substantial collection of viable goat egg cells. They obtained

---

* In 2009, the *Telegraph* (UK) ran an article titled "Extinct Ibex Is Resurrected by Cloning." The many subsequent news stories that reported these experiments erroneously gave 2009 as the year in which they took place. In fact there was a six-year delay between the successful experiment and the first scientific report in the journal *Theriogenology*, in 2009.

them by placing thirty domestic goats into a state of superovulation, using hormones to stimulate the ovaries to produce mature egg cells.

Working under a microscope and using micromanipulators, the experimenters removed the nucleus from each egg cell and replaced it with one of Celia's somatic cells. Next came the step that transformed the two cellular parts into a single working entity: the researchers applied two short pulses of electrical current to each cell, a process known as electrofusion. They then incubated the resulting fused cells, creating fifty-four reconstructed embryos. These were now essentially Celia's egg cells, as crafted through biotechnology.

Finally, according to Folch, "we transferred the cloned bucardo embryos to thirteen female recipients [who were either Spanish ibex goats or mixed hybrids], having two pregnancies that terminated spontaneously before day 75 of pregnancy."

All that toil and trouble resulted in no live births. So ended Experiment One.

In the winter of 2003, the researchers tried again with Experiment Two. This time they transferred 154 cloned embryonic bucardo cells into the wombs of 44 recipient goats. This yielded five pregnancies, one of which continued normally until term.

And then, according to the scientific publication describing the experiment, "at day 162 postfusion, we performed a caesarean section . . . One bucardo female weighing 2.6 kg [5.7 pounds] was obtained alive without external morphological abnormalities. The newborn displayed a normal cardiac rhythm as well as other vital signs at delivery (i.e., open eyes, mouth opening, legs and tongue movements) . . . To our knowledge, this is the first animal born from an extinct subspecies."

It was Wednesday, July 30, 2003, a turning point in the history of biology. For on that date, all at once, extinction was no longer forever.

ৡৡ   ৡৡ   ৡৡ

The Pleistocene epoch lasted from about 2.5 million years ago to about 10,000 years ago. That is getting awfully close to the present. The modern continents were about where they are today. But if ours is the age of global warming, the Pleistocene was the age of massive glaciation and global cooling. Glaciers covered as much as 30 percent of the earth's total land area, and in North America the ice sheet at one point extended as far south as what is now Chicago.

The Pleistocene witnessed the rise of the charismatic megafauna, animal species that included the woolly mammoth, Neanderthal man, and *Homo sapiens*. It also saw the extinction of many of them, including the mammoth and Neanderthal, although the reasons for the extinctions are unclear. While the advancing glacial ice depopulated the affected regions of plant and animal life, it did not necessarily destroy them: in many cases the animals retreated southward and continued to thrive. The wooly mammoth seems to have evolved in very cold habitats.

One explanation offered for the extinctions is that modern humans hunted many of the species into extinction. And they may have been at least partly responsible for the extinction of the Neanderthals, our genetically closest-related hominid species, as well as for the demise of the woolly mammoth. If this is true, the question arises whether we have an obligation to bring these creatures back, not as circus sideshow attractions but as part of a focused scientific attempt to increase genetic diversity by reintroducing their extinct genomes into the global gene pool.

The mammoth almost cries out for resurrection. Some specimens unearthed from permafrost are so lifelike that they appear to be merely sleeping, not dead, much less extinct. Consider, for example, the baby mammoth Dima, discovered in a northern Siberian gold mine in 1977.

Resurrecting such a beast looks almost easy, but when it was actually attempted, it proved to be anything but. In 1980 Viktor Mikhelson of the Leningrad Institute of Cytology tried to reconstruct a mammoth embryo from cells recovered from Dima, but gave up after several months of failure. In the case of Neanderthal man, moreover, we have only fossils, not cells, making a resurrection attempt even more challenging.

Neanderthal man was first discovered in August 1856, when miners working in a limestone quarry in the Neander Valley, near Düsseldorf, Germany, came upon a pile of bones. The workers thought they were the

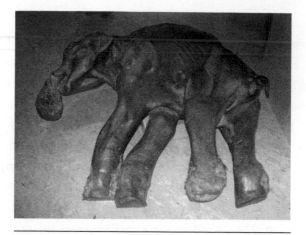

Figure 6.1 Dima

remains of a bear. In reality they were a partial skeleton of an ancient, lost species of humans.

Neanderthal man has since become a cultural icon, a fabled creature with a trademark name on the order of Godzilla or King Kong. But the Neanderthals were real, and for all the negative associations called up by their name, they were for a time the pinnacle of the animal kingdom.

They were bigger and stronger than modern humans, with larger skulls. And while "Neanderthal" has long been considered synonymous with "dumb brute," these people in fact manifested ample signs of reasonably high intelligence. They used stone and wood tools, as well as axes and spears. They applied body ornamentation, made fires, built relatively complex shelters, hunted and skinned animals, and ate meat. Neanderthals also buried their dead, sometimes together with flowers, a practice that suggests that they possessed some sort of primitive ideology or belief system.

Paleontologists have attributed these practices to Neanderthals on the basis of artifacts recovered from sites containing Neanderthal skeletal remains. Further knowledge of these people has come from an analysis of the Neanderthal genome. Scientists had long thought that the reconstruction of ancient DNA sequences was unlikely if not impossible, something that occurred only in the wilder reaches of Michael Crichton–style science fiction. The argument against it was based on the great age of the samples: any DNA recovered from ancient fossils would probably be too fragmented or corrupt to be readable. But in the 1980s Swedish paleontologist Svante Pääbo demolished that view once and for all.

Even as a child, Pääbo was fascinated by archeology and ancient civilizations, and at thirteen (in 1968) he persuaded his mother, who was a chemist (his father would win a Nobel Prize in Physiology or Medicine in 1982) to take him to Egypt. Later, while enrolled in a PhD program at the University of Uppsala, he obtained skin and bone samples from twenty-three mummies and hoped to extract DNA from them. To prevent contamination from human DNA, he did his work in an exceptionally clean laboratory, with a ventilation system sanitized by ultraviolet light. The lab was run in accordance with hot-zone–style biosafety procedures, with workers clad in sterile gloves, masks, and boots.

Pääbo ended up extracting and analyzing short stretches of DNA from the 2,400-year-old mummy of an infant boy. It was the first time in history that anyone had done such a thing, and his 1985 paper reporting his findings—his very first scientific paper, published while he was still a grad student—ran as a cover story in *Nature*: "Molecular Cloning of Ancient Egyptian Mummy DNA."

Pääbo then turned his attention to Neanderthal man. In 1997 he obtained from the Rhineland Museum in Bonn a half-inch (1 cm) bone sample of a 42,000-year-old Neanderthal fossil. He ground up the sample, dissolved it in a chemical solution, and extracted recognizable mitochondrial DNA fragments. Later, he and colleagues tested more than seventy Neanderthal tooth and bone samples and obtained useful DNA from six of them.

In May 2010 Pääbo and an international team published "A Draft Sequence of the Neanderthal Genome" in *Science*. Four of the gene sequences discovered by members of that team have brought us somewhat closer to a more accurate picture of Neanderthal man. Fragments of the MC1R gene suggest that the Neanderthals were likely to have light rather than dark skin. Another discovery was more portentous: the presence of parts of the *FOXP2* gene, which is involved in speech and language. This means that if we ever clone a Neanderthal into existence, we might actually be able to converse with him or her.

ə̂ɔ   ə̂ɔ   ə̂ɔ

But why resurrect a Neanderthal? Or for that matter, any other animal?

The most obvious reason for resurrecting extinct species is to attenuate, even partially, the wave of mass extinction that is currently taking place and is a hallmark of the Holocene—our own epoch. If the continuing loss of countless species is a tragedy, then the introduction of effective countermeasures, and the increase in species diversity that will accompany them, can only be viewed as a benefit.

If we can rescue one species from permanent extinction, then we can rescue others as well. Zoos are already becoming agents of species conservancy, keeping germ lines intact in living examples. In addition to living zoos there are also "frozen zoos," repositories of DNA as well as frozen, viable cell cultures, semen, embryos, oocytes, and ova, as well as blood and tissue specimens of extinct, rare, or endangered species. The Frozen Zoo at the San Diego Zoo, for example, houses samples from more than 8,400 individuals representing more than 800 species or subspecies. There is a smaller collection of genetic samples from rare and endangered species in cryopreservation at the American Museum of Natural History in New York.

Worldwide, there are about a dozen frozen zoos. The genetic material in storage there could be used in the same type of nuclear transfer cloning experiments that produced Dolly and the bucardo clone. Frozen zoos exist to amplify the gene pool, increase genetic diversity, and rescue populations of endangered species; successful efforts to revive species that are already extinct will have the same effect. Extinct species by definition have been removed from the gene pool. Cloning them back into existence will bring their lost genetic material back into circulation.

But how can cloning, which produces only carbon copies, be used to increase genetic diversity? If you clone from frozen tissues of dead animals, cloning them back into existence reintroduces their lost genetic material into the global population.

Some organizations are doing this systematically, for example, the Audubon Center for Research on Endangered Species (ACRES) near New Orleans. Established in 1997 with $20 million in public and private financing, its long-term goal is to "unlock the secrets that could make extinction extinct." Its immediate goal is to use nuclear transfer cloning to increase the gene pool, genetic diversity, and populations of endangered species.

ACRES researchers have been enormously successful, cloning several examples of a species of endangered African wildcat. But then they did something new: they bred together some of their clones, which then gave birth naturally. This was first done in the summer of 2005; the cloned wildcats produced two litters totaling eight wildcat kittens, all of them through natural birth. ACRES researchers thus have established that cloned animals can mate with each other and produce natural offspring. This creates additional specimens of the endangered animals, specimens that contain entirely new combinations of genetic material. Using these methods, ACRES has been able to increase the population of endangered Mississippi sandhill cranes by more than 20 percent in two years.

The director of the center, Betsy Dresser, makes several good points with respect to the benefits of cloning endangered species:

- Cloning can help eliminate disease in a population by cloning only the disease-free animals.
- Cloning versus saving the habitat is a false choice. You need to do both. Cloning provides a safety net.
- Cloning members of endangered species can help preserve and propagate species that reproduce poorly in captivity.
- Cloning can introduce new genes back into the gene pool of species that have few remaining members.
- Clones of healthy animals can be introduced into wild populations to give a "booster shot" to a species undergoing a loss of genetic diversity.

In addition, there are other practical reasons for regenerating lost species, and in particular for regenerating Neanderthal man. For one thing, the reintroduction of Neanderthals would give *Homo sapiens* a sibling species that would allow us to see ourselves in new ways. It might give us an inkling into another form of human intelligence, or of different ways of thinking. There might even be health benefits if Neanderthals proved to be resistant to diseases like AIDS or tuberculosis, for example, or diseases that coevolved with *Homo sapiens* like smallpox, polio, syphilis, or the next surprise pandemic.

Of course, there are also arguments against extinction reversal. An organism that has been extinct for 30,000 years is more likely to have little or no resistance to diseases that have evolved since then than to have a native resistance to them. Still, as we have seen, the human immune system offers little or no resistance to many of the diseases that our species has coevolved with, and a resurrected Neanderthal might be no worse off than we are in this respect. As a precautionary measure, newly regenerated species could be confined inside sterile environments until their disease resistance was evaluated and perhaps augmented through drugs, vaccines, or other modalities.

Another argument against reviving extinct species is that cloning is hard on the subjects, with any eventual successes being preceded by a long series of failed attempts: stillbirths, as well as misshapen, abnormal, and impaired offspring. Why bring these animals back only to have them suffer in this way?

There are at least two answers to this question. The first is that by the time regenerating these animals becomes economically feasible, cloning technology will have progressed to the point that successes will be far more common than failures. The second is that although nuclear transfer cloning may be hard on the animal, so is natural biological birth. In fact, approximately one in every thirty-three babies born in the United States each year suffers from one or more birth defects, which are the leading cause of infant mortality, accounting for more than 20 percent of all infant deaths annually. Birth is an inherently risky business.

More generally, anything can be done ineptly or expertly, carelessly or carefully, inhumanely or compassionately. This is equally true of attempts to bring back extinct species. There is no reason to think that extinction reversal will be carried out any less humanely than any other medical or experimental procedure. Indeed, the protocols already in place for the humane treatment of experimental subjects, whether human or nonhuman, can and ought to be extended to members of species that we might choose to bring back.

A final argument against extinction reversal is that to bring species back selectively, according to our own tastes and prejudices, will result in an anthropomorphized, "boutique" environment that reflects human values and judgments and which will result in an artificial construct rather than a natural phenomenon. However, we already live in such a world and have done so ever since the beginning of agriculture, if not long before. Everything from skyscrapers to golf courses, poodles to the Panama Canal, the Hoover Dam, Venice, and Las Vegas all attest to the fact that humans remake nature according to their wishes.*

Nor are we alone in this respect. Many other animal species also reconstruct the world according to their own wants and needs: birds build nests, beavers build dams, bees build hives, spiders spin webs and cocoons, and ants construct mounds, entire cities, and even ant cemeteries. A propensity for redesigning nature seems to be an inherent part of life itself.

All of that said, how do we bring back animals that have long since vanished from the scene?

❧   ❧   ❧

That depends on the species and what there is left of it. For species whose intact cells, tissues, or other genetic materials have been preserved—as in the case of the bucardo, for example, or species well represented in frozen zoos, labs, or other storage facilities—revival will be possible through straightforward interspecies nuclear transfer cloning.

A second category consists of species whose genetic material might or might not be too corrupt for cloning. This is true of the woolly mammoth,

---

* In 2000, the Nobel Prize–winning atmospheric chemist Paul Crutzen coined the term "Anthropocene" to refer to the geologic era in which human activity has had a significant impact on the face of nature.

for example. Although mammoth specimens buried in permafrost may look remarkably lifelike, their DNA is not in the same condition. The woolly mammoth, however, really falls into a class by itself.

Unlike the Neanderthals, whose remains consist exclusively of teeth, skulls, and other bones, Siberian mammoth carcasses have turned up with hair and even soft tissue on them. In August 1799 a Russian hunter looking for mammoth tusks came across an oddly shaped block of ice on the shoreline of the Laptev Sea in north-central Siberia. The summer sun had melted some of the ice, exposing two projections that later turned out to be tusks.

Two years later, the hunter returned to the area during the summertime and found that one side of the animal had been exposed to view, while the rest of the body was still frozen. It was not until 1806 that Michael Adams, of the Imperial Academy of Science at St. Petersburg, reached the site. By this point, nearby villagers had hacked off some of the flesh and fed it to their dogs. Bears, wolves, and foxes had eaten the rest, leaving only a skeleton. Now known as the Adams mammoth, the skeleton is on display in the St. Petersburg Zoological Museum.

More recently, the Jarkov mammoth was discovered in 1997 by a family of that name who came across a tusk protruding from the frozen ground of the Taymyr peninsula in northernmost Siberia. A group of latter-day mammoth hunters arrived at the scene and speculated that an intact mammoth carcass could be lodged in the ice—an entire mammoth! In October 1999, a helicopter lifted a twenty-three-ton ice block with tusks protruding bizarrely from it up and out of the frozen tundra, and hauled it to an ice cave. There, as recorded by a Discovery Channel film crew, scientists began defrosting the remains with hair dryers.

The Jarkov mammoth turned out to be mostly bones, but even so, a bit of soft tissue remained. It looked like a strip of beef jerky.

Coincidentally, a second defrosting mammoth (the Hook mammoth) happened to be located nearby, and some of the expedition's researchers traveled to the site. One of them, Alexi Tikhonov, cut off a piece of what appeared to be mammoth muscle. Jokingly, he offered it to those present all of whom refused this morsel. And so, braced by a few shots of vodka, he took a bite himself. "It was awful," he said. "It tasted like meat left too long in the freezer." (Mammoth meat is so common in Siberia that fox trappers use it as bait.)

Any mammoth tissue that is fresh enough to eat might harbor intact DNA. Two Japanese scientists have plans for resurrecting the animals. One of them, Kazufumi Goto, proposes finding intact mammoth sperm cells that he will use to inseminate a female elephant to produce a mammoth-elephant hybrid.

Hybrids are usually sterile, but there are known exceptions; whether a mammoth hybrid would be sterile is currently unknown. Assuming it wasn't sterile, then by injecting additional mammoth DNA into the result-ing mammoth-elephant cross, you would get a second hybrid that was even more of a mammoth than an elephant. According to Goto, repetition of the process with successive new offspring would yield, within fifty years, an animal that was 88 percent mammoth. (Elephants have a twenty-two-month pregnancy and don't produce offspring until they are ten.)

While this scenario is technically feasible, regenerating a woolly mam-moth in this way is a speculative possibility at best because intact mam-moth sperm cells have never been found—and might never be. The second Japanese researcher, Akira Iritani, is chairman of the Department of Ge-netic Engineering at Kinki University, near Osaka. He plans to find a mam-moth cell with intact chromosomes and then fuse them with an egg cell from an Asian elephant and let it divide to an early embryonic stage. Fi-nally he plans to implant the embryo into the womb of an Asian elephant and hope for the best.*

Richard Stone, who wrote a book about resurrecting mammoths, calls this "the Mount Everest of biology experiments." Everything hinges on lo-cating a fresh supply of exceptionally well preserved frozen mammoth tis-sue, which scientists would then defrost under carefully controlled laboratory conditions. This combination of circumstances has not occurred

---

* In August 2011 a mammoth thigh bone containing bone marrow was discovered in permafrost soil in Siberia. Kinki University scientists plan to use this material for cloning.

to date, despite the fact that mammoth hunters have conducted several field searches. Russian mammoth expert Sergey Zimov takes a dim view of these searches. "Frozen mammoths find you, not the opposite," he says. "Directed searches have almost no chance of success."

ߦ ߦ ߦ

A third class of extinct species is represented by DNA samples that are so fragmented and corrupt that their genomes must be laboriously reconstructed from innumerable isolated pieces. This is true of Neanderthal man, whose draft genome Svante Pääbo reconstructed in that manner. Unfortunately, the draft genome doesn't exist physically as actual chromosomes or genes, but only as strings of DNA sequences stored in computers.

Theoretically it is possible to convert those sequences into a physical, real-life genome by synthesizing short sequences (oligos) in DNA synthesis machines and then stitching them together into chromosomes. In 2010 Craig Venter created his so-called synthetic *Mycoplasma* bacterium by chemically synthesizing its entire genome, oligo by oligo. However, there is a huge difference between synthesizing a bacterial genome and synthesizing the genome of an animal as large and complex as Neanderthal man. While Venter's *Mycoplasma* genome was 1.08 million base pairs in length, the Neanderthal's genome consists of 3 billion base pairs, as long as that of a modern human. Synthesizing such an object oligo by oligo would take forever—or at least a very long time.

Fortunately, there's another way to accomplish the same objective: start with a physical genome that closely resembles the Neanderthal's and then change it, piecemeal, into the genome of a Neanderthal. Reverse-engineer it into existence.

What genome closely resembles the Neanderthal's? The modern human genome. In fact, the genomic difference between a modern human and a Neanderthal is about threefold more than between one modern human and another—about 10 million base pairs. Indeed some Melanesian genomes consist of up to 8 percent more closely related to Neanderthal and Denisova genomes than to the African genomes and hence would make a slightly (500,000 base pairs) better starting point.* (The chimpanzee is the next closest contender with about 30 million differences.)

Millions of alterations is a lot, and making them is all but out of the question for traditional genetic engineering methods, which introduce modifications one at a time, serially. We generally try to bring down costs before we undertake a large project (as happened with human genome sequencing). Recent work in my Harvard lab shows that you can reduce the costs of the process considerably by introducing the necessary changes on a batch basis, modifying multiple genetic sites at a time, and in parallel. This is the MAGE method, introduced in Chapter 3. One way to use it here would be to break up the human genome into 10,000 pieces of 300,000 base pairs each, and then replicate them in *E. coli* as bacterial artificial chromosomes (BACs) or in yeast as yeast artificial chromosomes (YACs). Then each piece can be reprogrammed in parallel by MAGE (about 1,000 changes each). By using chip syntheses, 10 million oligos can be printed at a cost of about $5,000 and at least double that to get them amplified and error-corrected in appropriate subpools. The twenty-three human chromosomes could be reconstructed in parallel (about 500 steps each) and then combined by chromosome transfer using cell or microcell fusion methods and multiple positive and negative selection markers. An example of a positive selection employs a drug resistance gene like neomycin phospho-transferase. When this resistance gene is attached to the BAC and exposed to a cell, then only the cells that take up the BAC will survive in the presence of the drug. Later, if you want to remove the piece of DNA that was needed temporarily, you can use a negative selection. An example used widely in mammalian genetics is a viral thymidine kinase gene not normally found in human cells that makes them sensitive to an antiviral drug.

Supposing, then, that we have recreated the physical genome of Neanderthal man in a stem cell, the next step would be to place it inside a

---

* Denisovans are another recently discovered species of "archaic humans." (They were named after the Denisova Cave in Russia, where bone fragments were found in 2008.) There is evidence that both Neanderthals and Denisovans interbred with modern humans.

human (or chimpanzee) embryo, and then implant that cell into the uterus of an extraordinarily adventurous human female—or alternatively into the uterus of a chimpanzee. Admittedly, this will only ever happen if human cloning becomes safe and is widely used and if the possible advantages of having one or many Neanderthal children are expected to outweigh the risks.

This same technique could be applied to the wooly mammoth once its genome was fully sequenced. In 2008 a scientific team headed by Stephen Schuster and Webb Miller of Pennsylvania State University reported in *Nature* that they had reconstructed a substantially complete draft sequence from clumps of mammoth hair. But that was sufficient for them to calculate that the mammoth genome differed at 400,000 sites from the genome of the African elephant. This shows us what the road to regenerating the wooly mammoth is.

You would begin with an intact African or Asian elephant genome (both animals are phylogenetically close to the mammoth), and then by using MAGE technology you'd introduce the modifications that would turn it into a mammoth genome. Finally you would implant that genome into an elephant's embryonic cell in the now familiar way, and implant it into the womb.

And then, twenty-two months later . . .

૪૨   ૪૨   ૪૨

A common objection to the idea of resurrecting extinct species is that since many of them disappeared due to the loss of their native habitat, it is pointless to bring them back into a world in which those habitats have long since vanished. But it is possible to bring back the habitat along with the animal itself. In the case of the wooly mammoth, there has already been an attempt at this.

In 1989 Sergey Zimov, director of the Northeast Science Station in Cherskii, Russia, together with a number of partners, established a nature preserve covering a sixty-square-mile area (about three times the size of Manhattan) that they called Pleistocene Park. Although it was in Siberia, within one hundred miles of the Arctic Circle, Zimov did not claim that the park, as it stood, reproduced the mammoth habitat of the Pleistocene epoch. The group's goal, however, was to turn it into one. "In some places we must not only preserve nature, we have to reconstruct it," said Zimov.

Zimov's hypothesis is that it was human hunting, and not climate change, that destroyed the mammoth. In his paper in *Science*, "Pleistocene Park: Return of the Mammoth's Ecosystem" (2005), Zimov argued that it was the animals themselves, more than temperature alone, that maintained the ecosystem in which mammoths thrived.

"It might not have been the climate changes that killed off these great animals and their ecosystem," he wrote. "More consequential, perhaps, were shifts in ecological dynamics wrought by people who relied on increasingly efficient hunting practices, which decimated the very population of grazing animals that maintained the tundra steppe."

Tundra steppe, the mammoth's primary habitat, was mainly grassland, the kind of open prairie that is common in the Midwest. But about 10,000 years ago, at the beginning of the Holocene, the mammoth tundra steppes of northern Siberia disappeared completely, replaced by an ecosystem rich with mosses and forests instead of grasses. According to Zimov, the loss of the tundra steppes was due to the loss of the mammoth, whose grazing habits had formerly kept the grasslands alive and fertile.

To return the area into a mammoth ecosystem, Zimov suggested introducing large herbivores selectively. To that end, in the spring of 1998 Zimov brought to the park thirty-two Yakutian horses, the breed that was closest to those that lived in the region during the Pleistocene. Over the following three years the horses converted a confined area of the park from mosses and shrubs to grassland. Later, Zimov imported some two dozen wood bison from Canada. Gradually, a large, fenced area of Pleistocene Park came to resemble the mammoth ecosystem that was in place when the last mammoths roamed the earth.

If and when woolly mammoths are ever cloned into existence, bringing them to Pleistocene Park would be a case of returning them to their natural habitat. It would be the closest thing to time travel: a return to the flora and fauna of the Pleistocene epoch, a sort of latter-day Siberian Eden.

It would also turn the area into an adventure tourist destination, for the park would in effect be a mammoth zoo.

It might be a while before that happens. The first animal to be resurrected from extinction, the clone of Celia, the bucardo, lived for about seven minutes before dying of a lung condition common to cloned mammals.* Seven minutes might not seem like much, but then the first flight of the Wright brothers in December 1903 lasted for all of twelve seconds. Sixty-six years later, in 1969, we were on the moon.

---

* Robert Lanza, of Advanced Cell Technology, an adviser to the bucardo project, speculated that if a pulmonary surfactant, which aids breathing, had been promptly and properly administered, the animal might have lived.

CHAPTER 7

-10,000 YR, NEOLITHIC

Industrial Revolutions.
The Agricultural Revolution and Synthetic Genomics.
The BioFab Manifesto

ge

Industrial Revolutions

The Neolithic era began roughly 10,000 years ago in the Middle East, at the tail end of the Stone Age. By this time in history the genus *Homo* had winnowed itself down to only one remaining human species, *Homo sapiens*—us—which had by then vanquished, assimilated, or otherwise outsurvived Neanderthal man, the Denisovans, and all earlier examples of archaic humankind. The period was put on the map and immortalized by the development and use of polished stone implements—and nice-looking ones at that—as opposed to the chipped or found stone tools utilized in the earlier Paleolithic.

But the Neolithic is noted for something far more important than stone tools—the invention of agriculture. Other than the massive set of effects wrought by the industrial revolution, the single greatest transformation in human history occurred during the Neolithic, as people turned from hunting and gathering to farming and animal husbandry.

The agricultural and other industrial revolutions are major turning points in human history because they allowed us to make immense leaps in our understanding of and control over nature. They were revolutions in knowledge and in toolmaking, and they have clear analogs in synthetic biology, which is likewise a product of specialized knowledge and a unique set of tools.

Human history includes at least six different "industrial revolutions." Arguably we are now in the midst of the sixth industrial revolution, and the tools and knowledge it encompasses have given us the power to remake ourselves. Revolutions are sometimes scary, but they do not have to be. Each revolution begins with a period of tinkering by trial and error. A prehistoric "scientist" stumbles across a fire and tries adding dry leaves that start to burn, hot as the sun. Then he tries adding sand but discovers that this puts the fire out. Revolutions spread outward from the center with vague ways of communicating intentions and degrees of progress, as when our fire man tries to tell his friends that fire is hot, and that friend tells others. Eventually we develop measurements, in this case scalar indications of temperature, and models that enable prediction and design.

Revolutions can have unanticipated positive and negative consequences—as when a fire rages out of control and perhaps incinerates its maker. Bearing this in mind, we will chart the course of the revolutions that have led to the power to control our future biological development—to understand and then manipulate the evolving genome of life itself.

The first industrial revolution was centered on the notion of time. It began 15,000 years ago, when we had no idea of what 15,000 years meant or what a revolution was. Those of us who looked human (including Neanderthals and Denisovans) had spread far beyond Africa and were discovering a need to understand time, so that we could predict the seasons. We could get by in the transition from the season of gathering foods to the season of planting them just by waiting for the warmth of spring. Why, then, bother measuring time?

Because it was reassuring during the winter to visualize the time remaining until spring to pace the use of stored food. Floods and droughts recurred, somewhat predictably, each year. Some life-altering events took place on longer time frames like the five-year cycle of El Niño. Some catastrophes occurred less frequently and lacked a periodic component but required a collective memory to maintain preparedness. Fortunately, the crucial measurements tended to be easy and digital and could be checked against other measures. Thirty sunrises corresponded to one lunar cycle. Twelve lunar cycles made up a year. The day wasn't digital but rather smoothly analog, and dividing it into hours with a sundial and precise seconds with a mechanical clock probably wasn't crucial until we started serious navigation.

What was the killer app, or tool, for measuring time? In terms of tempo, biological systems exhibit natural cycles that are synchronized with some astrophysical cycles. Biological cycles such as times of hibernating, mating, and flowering match up with the earth's tilted revolution around the sun. Bears wake up slowly at the end of winter, and then their prey animals get a fast and rude awakening as the season's first bear claws penetrate into their resting places. Matching the lunar cycles most evidently are the tidal behaviors. Less obviously, some animals (e.g., primates) menstruate monthly, while other mammals generally have nonmonthly estrous cycles. Matching the rotation of our home planet, almost all life has circadian, diurnal cycles of metabolism. Probably all animals with brains have tendencies toward sleep patterns synchronized to the sun. Cave-dwelling animals lost this synchronization over the course of many millennia.

At the scale of seconds, we notice heartbeats and wing beats. In the millisecond range, whales and bats produce ultrasound vibrations of 100,000 per second and up to 180 decibels to navigate and communicate. Some biological systems purposefully avoid simple patterns in order to thwart predation; for example, seventeen-year cicadas and eighty-year cycles for the blooming of bamboo. The point is that the first clocks were hardwired into living things of all stripes, and then human beings started reinventing them and soft-wired them into our culture. Initially this was in service to the gods of agriculture but the study and engineering of time spread aggressively into many of our technologies today. Our close relatives the great apes tend to think on very short timeframes—instant gratification. The ability to tell long narratives in the form of epic poems and songs, and to draw cave paintings (as far back as 32,000 BCE), went hand in hand with a growing awareness of causality and the advantages that such awareness brings. This contributed to developing strategies for hunts and for warfare that required more coordination and timing than even the remarkable skills of wolf packs.

Keep in mind that it doesn't take much of an advantage for a revolutionary advance to sweep through a population. A 5 percent advantage compounded annually for twenty years is a 260 percent advantage, and over two hundred years is a 17,000-fold advantage.

As with most technologies, the taming of time bore unwelcome and unanticipated consequences. Today, as we face impending deadlines, a hectic pace of life, and existential risks of all kinds, it's tempting to think that stress was less severe in prehistoric times. But we have had many generations to adapt evolutionarily, while the revolutionary concepts of time and causality may have had a comparatively rapid onset. The unwelcome consequences of warfare and stealth and deception reverberate in our culture and inherited psyche today.

The moral of the story is that progress comes with hidden costs, risks, and unpleasant surprises. As I chart the course of genomic technologies, I will do my best to point them out.

The Second Industrial Revolution, 4000 BCE:
The Agricultural Revolution and Synthetic Genomics

Agriculture, the domestication of animals and crops, and the trade it resulted in encouraged the concentration of people and led to cities. Probably the first domesticated crop was emmer wheat (*Triticum dicoccoides*), found growing wild in the ancient Near East. Two wild grasses, *Triticumurartu* and *Aegilops speltoides*, had intergenus sex. They were diploid (2X), meaning that they had one copy of each chromosome from their mother and father, but their intergenetic children were tetraploid (4X), meaning that they kept two copies each—a full set from all four grandparents. This is rare in any given generation but common over evolutionary time, and ranges from triploid (3X) watermelons and water bears to dodecaploid (12X) plumed cockscomb (*Celosia argentea*) and clawed frogs (*Xenopus ruwenzoriensis*). The tetraploid wheat hybrids were adopted by humans possibly as early as 17,000 BCE (based on carbon-14 isotopic dating), in what is now southern Turkey (based on DNA studies), and then spread as far as Egypt to feed the pharaonic dynasties. Along the Yangtze River we see another dramatic domestication process dating from 12,000 BCE: changes in the morphology of rice phytoliths. And yet another in the Balsas River valley in southeastern Mexico around 6700 BCE, when an annual grass, *Zea mays,* began its long transformation into modern corn. Domestication of thousands of additional species of plants and animals followed.

Possibly predating agriculture were dense populations of animals, minerals, or vegetables that hunter-gatherer tribes concentrated and then began to trade and protect their resources. Advantages of concentrating people and their material goods included the scaling of construction—for example, the number of people within a walled enclosure goes up with the square of the material to build the wall. As the density of such wealth grew, so did civil engineering of buildings, walls, boats, and bridges. This required the invention of measurements of length, weight, and cost. The consequences of poor measurements could be fatal. The misalignment of a wall could result in the collapse of a building when tested by a storm or invaders. Ancient architects are said to have been required to stand under their arches when first load-tested. More recently, in 1999, the $328 million Mars climate orbiter mission failed due to the use of incorrect units of force (pound-force vs. newton).

The unwelcome consequences of the second industrial revolution and the resulting abundance included diseases of crowding. Cholera is caused by the gut bacterium *Vibrio cholerae* found in contaminated drinking water. The chance of such contamination rises sharply with the density of people and is fairly rare in other animals. Similarly, *Yersinia pestis*, the causative agent of black plague, depends on high concentrations of grain, which bring rats, which in turn bring the fleas that harbor the plague organism. This phenomenon is often associated with two waves of plague, the Black Death, which spread from China to Europe between 1330 and 1360 CE. The first documented instance of a plague epidemic occurred in 1200 BCE, around the time that the Philistines stole the Ark of the Covenant from the Israelites and then returned it (possibly thought of as a means of escaping a curse). A recurrence of the plague in 540 CE, in Ethiopia or central Asia via Egypt, spread by ships and caravans of the Emperor Justinian, killed as many as 100 million people. In the Middle Ages another 75 million died (roughly 50 percent of many European villages). Another 10 million died in Asia in 1885 CE. Gabriele de'Mussi's contemporary account of the 1346 siege of the Crimean city of Caffa (currently Feodosia, Ukraine) describes Mongol soldiers catapulting plague-ridden Mongol corpses into the double-walled city, and constitutes one of the first documented instances of biological warfare.

Malaria arose from expanses of stagnant water in rice paddies and other irrigated crops. Celiac disease (a failure to digest food caused by a hypersensitivity to gluten in the small intestine) arose when wheat became plentiful in our diet before our genome had a chance to adapt—or more accurately stated, before those adaptations had spread to all wheat eaters. The convenience of monoculture crops brought with it monoculture pests, like locusts. Plowing removed meter-thick roots that fought erosion. The use of plants lacking nitrogen-fixing bacterial ecosystems resulted in soil depletion and the need to fertilize. Fertilization, in turn, resulted in runoff into ponds with consequent blooms of microbes, which consume so much oxygen that fish can't survive. So they go belly up.

❧   ❧   ❧

The switch to agriculture had several further consequences. Whereas hunter-gatherers existed in small, mobile, roving bands, early farmers lived near their fields in order to protect them from predators and plunderers, as well as to harvest and process crops. Harvesting, in turn, required the development and use of new tools and implements such as plows, sickles, and milling and grinding stones. Houses, community centers, and then villages, towns, and cities arose near these fertile areas. These city inhabitants led more sedentary lives than their hunting and gathering forebears. Social life became more complex, structured, and hierarchical than ever before.

Farmers often grew more food than they could use, which led them to develop storage vessels, bins, and storehouses. More important, the accumulations of foodstuffs prompted the early farmers to trade with other people, which helped create a working economy and led to new concentrations of wealth.

Further, whereas hunter-gatherers tended to exhaust the resources of a given region and then move on to the next, only to despoil it, the early farmers actually improved and increased the yield of a given piece of land through cultivation and irrigation. Instead of merely letting wild plants resow themselves wherever their seeds happened to fall, the farmers preferentially sowed seeds of plant types that were hardier, bore more fruit, were better looking, tastier, or otherwise viewed as more desirable than lesser species. This was a form of artificial selection, a favoring of one sort of plant type over another, and of increasing the numbers of the favored plant at the expense of those considered to be less attractive.

The domestication of plants and animals that occurred during the Neolithic era has clear parallels to synthetic biology—the attempt to domesticate microbial, plant, and animal genomes, including those of humans. Synthetic biology has progressed in three distinct phases. The first was the era of genetic engineering or basic biotechnology. Starting in the 1970s, this was the time in which researchers "domesticated" microorganisms by modifying their genomes manually. They first got *E. coli* to produce insulin, erythropoietin, monoclonal antibodies, and other such substances. The tools they used to modify genomes, while seemingly advanced for their time, are nowadays viewed as more or less Stone Age devices.

The second phase of synthetic biology is a period of growth and elaboration, with commercial synthetic genomics extended to a wider set of goals such as the discovery and production of new drugs, biofuels, and

genetically modified foods. It's also characterized by the use of more so-
phisticated tools in the form of automated techniques and machines, and
the development of novel methodologies such as the use of induced
pluripotent stem cells to create a range of narrowly targeted pharmaceuticals. Doing all this successfully on a mass, industrial scale further required the invention of implements that are comparable in their way to
those developed in the mechanization and industrialization of agriculture
by means of tractors, harvesters, threshers, combines, automatic cow-milk-
ing machines, and the like. In both instances, it was the age of commercial
mass production of the respective commodities.

A third phase of synthetic genomic enterprises is now in the making.
These commercial enterprises will try to make a living out of synthesizing
entire new genomes. At first glance, this may seem like an unprecedented
and entirely novel development. But in actual fact, this advance, like the
others we have considered, only recapitulates what nature had already
done. Nature, after all, was the pioneer genome synthesizer. Nevertheless,
if and when we can duplicate what nature has done and create a new
genome with never before seen functionality from scratch, then we might
finally be in a position to claim that we really know and understand how
life works, from the molecules up.

ᵯ   ᵯ   ᵯ

The immense impact that agriculture had on social complexity is paral-
leled by the impact of molecular technologies on the life science industry.
In the Neolithic, the simplicity of spears and fire gave way to oxen pulling
wheeled carts on roads, to food and seed storage, irrigation, and so on. In
like manner, the simple elegance of the dawn of molecular biology can be
seen in our ability to sketch out the essentials of the genetic code merely
by looking at phage plaques on a petri plate and then binding oligoribonu-
cleotides to ribosomes. That has changed beyond recognition.

Biotech research has since followed an amusing downhill path of
"progress." In the early days, organismal and molecular biologists would
compete with each other to enact hands-on feats of bravado, toughness,

and self-reliance. They would hunt and gather in extreme, hostile locales (from snake-ridden jungles to biosafety fume hoods full of radioactive mutagens); they would stop whirring centrifuge rotors with their bare hands, make their own enzymes, and work with high levels of radioisotopes straight out of a nuclear reactor. Toiling for a couple weeks to make a single enzyme, however, people soon realized that it was just as easy to make enough to last them and all their friends a year as to make a batch that would last one person for a week. In the mid-1970s this realization prompted small companies, such as New England Biolabs and New England Nuclear, to make enzymes used for recombinant DNA and for labeling reagents. That development was similar to an earlier trend for making and supplying long lists of chemicals to researchers, for example, by Sigma and Aldrich (which merged in 1975).

So the next generation of biology students got lazier and less in touch with the basics underlying the synthesis of the enzymes and the chemicals and reagents; instead of making them, they purchased them (while the old-timers moaned). The next step in the devolution was the idea of kits. Researchers found that they had large collections of expensive stuff that failed to perform as expected. This could have been due to generally bad protocols and lack of training, for example. But the solution that appealed to companies was the idea of selling kits—a set of enzymes and chemicals (often including an exotic ingredient known as water) that were individually necessary and jointly sufficient for success at a common lab task, and all of it was attractively packaged and presented.

Big hit! Researchers suddenly became more productive, which meant more sales . . . and eventually repetitive stress syndrome.

Another next step in the evolutionary degeneration of lab researchers was "instruments." The kits could lead researchers by the hand, as it were, but human errors were still possible: why read that fat manual anyway? The solution this time was to take the human out of the loop. Translate the manual into robot code. And while we're at it, have the robot use ninety-six pipettes instead of just one, so that ever more enzymes, chemicals, and reagents are needed. Lewis Carroll's Red Queen comes to mind: "Now, *here*, you see, it takes all the running you can do, to keep in the same place. If you want to get somewhere else, you must run at least twice as fast as that!"

But then a proliferation of competing instruments, combined with high capital costs, steep learning curves, downtime, and rapid obsolescence led to even higher levels of stress than before. This time the inevitable solution was "services." Centralized core facilities would own and operate the expensive machines and employ professionals (the priesthood) to maintain and upgrade the machines. This led to a deluge of data (and a need for bigger grants). But how would we handle all of this data? By building databases. Finally someone noticed that building a database wasn't the same as interpreting the data, and so we now have informatics services, cloud computing, and whatever else is out there on the horizon.

So in summary, the descent of man (the devolution of research persons) went like this: (1) DIY. (2) Buy parts. (3) Buy kits. (4) Buy machines. (5) Buy services. (6) Buy interpretation. This was "progress."

The BioFab Manifesto

As noted earlier, commercial synthetic genomics needed its own new set of tools, and so in 2004, nine biotechnology-minded scientists, including Drew Endy, Jay Keasling, and myself, among others, banded together to start an informal organization—the BioFab Group. By emulating the engineering practices that made silicon chip technology so successful, we hoped to make biological engineering into an equally revolutionary enterprise. What had hitherto been called genetic engineering, we were convinced, was not really engineering in the true sense of the word. Engineers normally had access to an ordered supply of well-defined, interchangeable, off-the-shelf parts, specification sheets, system plans, and so on. Genetic engineering, by contrast, worked largely by hit or miss, trial and error methods. It was more like an artisanal enterprise, or a craft, than an engineering discipline worthy of the name.

In 2005 the nine BioFab co-conspirators wrote a paper—a sort of BioFab manifesto—describing how we intended to transform biological engineering from a craft into an industry. It appeared in 2006 as a *Scientific*

American cover story: "Engineering Life: Building a FAB for Biology." Many of the techniques and devices, principles and practices we proposed there are now staples of the second generation of commercial synthetic genomics.

In general, fundamental engineering principles include (1) the use of interoperable parts (e.g., USB ports and plugs, dual inline pins for chips); (2) standards (e.g., maximum variation in properties, stress resistance); (3) hierarchical functional designs (e.g., designing the windshield without worrying about the hubcap design); (4) computer-aided design (CAD); (5) cost-effectiveness (e.g., miniaturization of integrated circuits, assembly lines); (6) physical isolation (e.g., car windows, plastic-coated insulated electric wires); (7) functional isolation (e.g., car keys, FCC regulations on radio emissions from circuits); (8) redundant systems (e.g., seat belts plus roll bars and air bags); (9) ruggedness and safety testing (e.g., crash dummies); (10) user licensing (e.g., driver and car registration); (11) user surveillance (e.g., radar monitoring of individual vehicle speed, stoplight cameras); (12) functional integration (e.g., dashboard clustering of instrumentation); (13) evolution (e.g., field testing and market feedback). Items 6–11 are also relevant to safety and security.

The biological equivalents to the principles and examples mentioned above are (1) interoperable parts (e.g., biobricks; see Chapter 8); (2) standards (e.g., protocols for connecting one biological part to the next; see Chapter 8); (3) hierarchical functional designs (e.g., designing transcription termination sites levels without worrying about how they start); (4) computer-aided design (CAD) (e.g., caDNAno—software for the design of three-dimensional DNA origami structures); (5) cost-effectiveness (e.g., choosing organisms that replicate quickly and do not require lots of expensive nutrients, land, energy, or other scarce resources); (6) physical isolation (e.g., sterilizable bioreactors, biosafety level BSL4 labs, and moon suits); (7) functional isolation (e.g., new genetic codes; see Chapters 3 and 5); (8) redundant systems (e.g., nutritional constraints plus isolation; item 6); (9) ruggedness and safety testing (e.g., building radiation resistance into organisms, testing genomically modified organisms in biosafety labs before releasing them into the environment); (10) user licensing (e.g., registration of DNA synthesis machines, chemicals, and lab personnel); (11) user surveillance (e.g., requiring commercial DNA synthesis firms to check orders against known pathogen sequences); (12) functional integration (e.g., codesigning photo-bioreactors and cyanobacterial genomes); (13) evolution (e.g., MAGE, sensor selectors). (For items 6–13, see the Epilogue.)

Item 13, evolution, differs vastly between biological and other types of engineering. In nonbiological engineering, huge resources go into developing a new product, and so there is rarely more than one or a few new products in a given niche per company, and only a few companies worldwide operating in those niches. In biological engineering, however, researchers can intelligently design billions of genomes and, given clever selection (see Chapter 3), one or a few winners will emerge from a lab-scale version of Darwinian survival of the fittest. And each year we get faster and better at laboratory versions of evolution.

Genome Engineering Becomes a Business

Just as the fundamental principles of engineering apply to biology, so basic business principles are being applied and adapted to the field of genomic engineering. The process of extending engineering principles to biology and applying business principles to genome engineering were logical and indeed inevitable developments. But at the beginning, these ideas were so counterintuitive that it took forty years from the dawn of molecular biology to the dawn of molecular biology engineering, and about twenty years to adjust to the business challenges and even to the philosophical and policy components of the new business models. Some business practices that are taken for granted in mechanical, chemical, civil, and aeronautical engineering took on strange twists when applied to human beings and form a type of "biological exceptionalism." By exceptionalism, I mean that an invention that is useful and nonobvious would be embraced if made from chemicals, but not if made from cells. What people debating the topic of "patenting life" sometimes forget is that the alternative to patenting biological inventions is a burgeoning of trade secrets. Truly creative or serendipitous breakthroughs will get hidden, possibly for years, until rediscovered independently. A patent is a limited monopoly to coax such secrets out into the open. Seldom have such monopolies impeded research. Occasionally there are rumors of intimidation—for example, labs making homebrew enzymes (Taq-polymerase, owned by Roche) for do-it-yourself gene amplification. The scariest examples of monopoly are tools for targeted DNA recombination (Cre-lox, owned by Dupont) and diagnostic breast cancer genetic diagnostics (*BRCA1* and *BRCA2,* owned by Myriad Genetics).These patent monopolies probably don't prevent research, but they may be perceived as preventing or reducing profits for new inventions building on the older ones.

The term "synthetic biology" originated in 1912 with Stéphane Leduc's work "La biologie synthétique, étude de biophysique," but has been embraced since 2004 to provide an intentional new feel to the enterprise, to distance ourselves from previous fields like genetic engineering, and to encourage a new rigor around the dozen or so fundamental engineering principles listed above. But the field soon splintered into six different camps: (1) recombinant DNA applied to metabolic engineering (Keasling), (2) bio-inspired and bio-mimetic (Steve Benner and Bob Langer), (3) engineering-inspired biobricks (Tom Knight), (4) evolution (Arnold, Ellington, Church), (5) instrumentation (the Wyss Institute and various companies), and (6) genome engineering (JCVI, HMS).

It makes a huge difference which camp you bet on. Critics might make the following objections: (1) Old-fashioned single gene bashing is low throughput. (2) Mimics don't take full advantage of real biology. (3) The ideas of modularity and biocircuits are naive, context-dependent, and not robust. (4) Historically evolution is not fast (relative to quarterly reports). (5) While the markets for bioproducts may be huge, the market for machines to improve production may be tiny. And as costs plummet, how do you expand this tiny market? (6) Why synthesize a whole genome if you only need to change a few bits?

Entrepreneurship is not for the faint of heart, and this is just as true for the synthetic genomics business as it is for any other. Companies, like technologies, tend to come and go. Indeed, the reading and writing of DNA probably constitute the fasting moving technology freefall in history (Figure

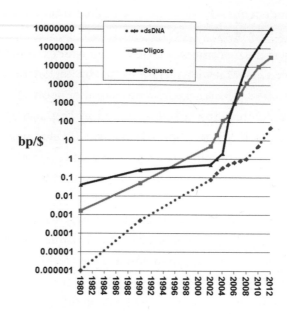

Figure 7.1 Exponential technologies. Gordon Moore's law for the density of transistors in large-scale integrated circuits has accurately held at a spectacular exponential 1.5-fold improvement per year for the cost of various aspects of electronics,
disk drives, telecommunications, and applications since the mid-1960s (or perhaps
all the way back to semaphores in the 1790s). Similarly, DNA reading and writing
technologies were on the same (1.5-fold per year) curve until 2005, when so-called
next generation technologies started to arrive and the exponential shifted to an almost tenfold increase per year. While in principle you can automate anything, au-
tomation alone doesn't guarantee rapid cost changes. The new (tenfold per year)
curves required a radical shift in methods from the use of columns through which
DNA flows to the use of flat glass on which multiplex immobilized DNA reactions
can occur and hence be downsized in volume and scaled up in numbers. We now
can multiplex billions of reading reactions in one flow cell and 30 million oligos
synthesized on one wafer (e.g., using Agilent ink-jet process).

7.1). There are multiple pathways ahead, and so companies can merge with
other companies. They can get acquired by other companies, and their
technology and people can be split up in various ways. With all that in
mind, we turn to a capsule history of genomic engineering as a business,
a terrible and awesome saga, complete with vignettes of major players,
personal asides, and even pithy apothegms.

Exploit the Fear Genie

You might think that the best way to obtain government funding is to lobby Congress or form a blue ribbon advisory committee at the august National Academy of Sciences or form a grassroots movement that gathers signatures. But then you would miss the swiftest and most powerful motivator of all: fear. In 1987, 1995, and 1998 three companies, working independently, received billions of dollars from the United States and the world for a previously unknown backwater of biomedical research: genome sequencing.

In the spring of 1987 Wally Gilbert, onetime CEO of the Cambridge biotechnology firm Biogen, decided that the time was ripe for a company to prospect and lay claim to as much of the human genome as possible, and swiftly. He used his Biogen golden parachute to recruit the pack that would later found the Human Genome Project (HGP)—Lee Hood, Sydney Brenner, Eric Lander, Charles Cantor, and myself—and announced the formation of a company with the working name Genome Corporation, for the purpose of sequencing the human genome and copyrighting and selling the data. (The name Terabase was used internally as a play on terabase meaning a trillion DNA bases and terrabase meaning a base for world domination.)

But the stock market crash of October 1987 made the capitalization of the Genome Corporation impossible, and so the company died on the operating table, as it were. Nevertheless, the public reaction to our plan to *sell human genomic information* (!) was speedy and alarmist: we do not want any company "owning" our genes, our priceless genetic heritage, our children's future. Congress appropriated $3 billion to NIH over the next fifteen years to sequence the human genome (a buck a base), and voted to make the information free to all humanity before any company could claim it. This was framed as a new budget line item to avoid criticism from entrenched NIH grantees that mediocre "big science" (huge academic genome centers) would steal from ongoing "small but great science" (the science of the small grantees, naturally).

The second fear genie was unleashed in 1992, ironically, when NIH lawyers insisted on patenting bits of human source code that Craig Venter's team had sequenced from RNA molecules. This got enough negative attention that a major investor, Health Care Investment Corporation, coughed up $70 million to form Venter's nonprofit TIGR (The Institute for Genomic Research) and the for-profit Human Genome Sciences (HGS). The genes TIGR sequenced and patented would be licensed to HGS for commercialization. That arrangement poured fuel on the flames, and the public launched a supplementary effort, the Merck Gene Index Project (MGIP), on top of the already huge Human Genome Project to directly compete with HGS. This supplementary effort involved companies that normally would not work together, but in this case were less threatened by each other than by a single company (HGS) owning most of the human genome. The reactionary tactic worked (sort of). During the ensuing fool's gold rush, the patents filed by TIGR did not generally add sufficient value to get far enough away from unpatentable "products of nature." Nevertheless the juggernaut of sequencing mRNA molecules would continue, especially once next-generation sequencing made it a viable way to quantitate RNA levels.

The third and largest genie appeared in 1998 when the PE (Perkin-Elmer) Corporation, later renamed Applera, created Celera. This was the private firm, successor to TIGR, where Venter used the shotgun sequencing method (initiated by Roger Staden in 1979 and applied to bacteria in 1995 at TIGR) to essentially speed-sequence the human genome. (Francis Collins, director of the competing public Human Genome Project, and no big fan of Venter's, said that shotgun sequencing would produce "the Cliffs Notes or the Mad magazine version" of the genome.) Venter's plan at Celera was to patent certain gene sequences and license them to commercial firms for a fee, while at the same time publishing unpatented sequences on its web page.

Coincidentally, Celera exhibited the newest model of Applera's sequencer (at the time known as the Applied Biosystems sequencer). This in turn drove sales of the sequencer to both sides in the human genome Olympics.

So who won? As Nietzsche once said, "All and none."

In June 2000 Venter and Collins, standing on either side of Bill Clinton, announced the completion of the draft human genome to the public. John Sulston, head of the small British genome project, said in his book about the HGP, *The Common Thread*, "The date of the announcement, 26 June, was picked because it happened to be free in both Bill Clinton's and Tony Blair's diaries. [There was a simultaneous UK announcement on the same date.] It was not clear that the Human Genome Project had quite got to its magic 90 percent mark by then, and Celera's data were invisible but known to be thin, so nobody was really ready to announce; but it became politically inescapable to do so. We just put together what we did have and wrapped it up in a nice way, and said it was done. . . . Yes, we were just a bunch of phonies!"

Celera (corporate motto: "Speed matters. Discovery can't wait.") survived, but didn't sell much of its original product (proprietary human genome data). The public probably benefited from an early end—even though the Clinton-Blair victory declaration was something of a sham. But the early end allowed us to move on to totally new methods.

The Sequencing Pandemic

By 2004 the sequencing market was feeding 13,000 of those Applera (née ABI) machines. The company maintained a monopoly on sequencing for about a decade, and had not exactly encouraged innovation, even among its users. Various users had figured out how to get more samples per machine, or better quality of automated reading, or recipes for homemade key enzymes like Taqpolymerase. Rumors spread that researchers had been sent threatening letters for some of these unauthorized applications, and that the instruments had been changed to discourage such "off-label use."

Once the genome war ended, the real fun began. Various alternative sequencing methods had been percolating for years in the shadows of the "race for the genome." There was the genomic/multiplex sequencing scheme that Wally and I published in 1984; sequencing by hybridization

(SBH) that Rade Drmanac and colleagues proposed in 1989 and demonstrated in 1993; polonyFiSSEQ (short for polymerase colony fluorescent in situ sequencing) in 1999; and single molecule sequencing methods championed in the early 1990s by Seq Ltd and Los Alamos. The NHGRI (the National Human Genome Research Institute) asked for input for roadmap projects to keep the genomics effort going as the HGP drew to a close, and a few of us suggested a technology goal, specifically a $1,000 human genome. Today we are actually closing in on that goal.

The sequencing field had progressed at about 1.5-fold exponential per year from 1970 to 2004, and from 2004 to 2011 accelerated to tenfold per year. Then we reached a tipping point where there was a veritable explosion, almost an epidemic, of new companies and divisions that collaborated with and/or sued each other. Here, for the record, is a list of them, fairly complete (so far as I know) as of April 2012. Already available technologies include six based on light microscopy of amplified clusters of molecules (Roche-454, CT; AB-SOLiD, MA; Dover Polonator, MA; Illumina, CA; Complete Genomics Inc., CA; Intelligent Bio, MA); two based on fluorescent light from single molecule (Helicos, MA; Pacific Bio, CA); and a nonoptic method (Ion Torrent, CT). Not yet available is a mix of some in R&D phase, some in the market but not for full genomes, and some abandoned (GnuBio, MA; Stratos Genomics, WA; Halcyon, CA; ZS Genetics, NH; Bionanomatrix, PA; LightSpeed, CA; GenizonBioSci, QC; LaserGen, TX; GE Global, NY; Genovoxx, Germany; Visigen/Starlight, TX; Genapsys, CA; Nanophotonics Biosci, CA; Base4innovation, UK; Mobious Genomics, UK; Reveo, NY). The six top nanopore sequencing contenders are Nabsys, RI; OxfordNanopore, UK; Electronic Biosciences, CA; IBM-Roche, NY; Genia, CA; NobleGen, MA. Nanopore technology has the questionable distinction of taking the longest time to get from idea to market starting with first patent filed in 1995 and on to the huge boost in February 2012 with the announcement at the AGBT meeting that OxfordNanopore had a handheld, disposable USB stick sequencer, called the MinION, capable of reading a billion base pairs, and a bigger sibling on the way capable of a complete human genome sequencing in fifteen minutes. (Nanopore sequencing works by measuring the rate that ions flowing through a pore like the one in Figure 2.1 changes as different portions of a single-stranded DNA sequence flow through it.)

When in Doubt, Outsource

In 1989 I met Craig Venter, who had invited me to the first genome sequencing meeting at the Wolf Trap conference center in Vienna, Virginia, just outside of Washington, DC. By then Craig had a brilliantly simple strategy for sequencing DNA. He was stuck at NIH, where researchers had loads of money but very little space, and great limitations on hiring people. So he brought human cDNA molecules (complementary DNAs useful for a variety of molecular-biological lab tasks) from plasmid libraries from one company (Clontech), and then sent them to another company to grow the plasmids and purify the DNA, then pay ABI (Applied Biosystems) to set up and maintain instruments that would sequence the pure plasmids. Finally he would send the data to Genbank to store in professional databases run by David Lipman and Jim Ostell, leaders in the development of cutting-edge search tools.

Thus Craig's little NIH team accomplished what other labs could not because they were struggling to develop ways to do all of these steps on their own and to minimize costs. Perhaps they felt that their peers would not fund or renew their grants if they outsourced nearly every aspect of their research, but Craig didn't buy into that penny-wise, pound-foolish dichotomy. The success of his approach surprised its numerous critics. When the project transitioned to TIGR-HGS, and NIH became concerned, I claimed that NIH shouldn't worry, since completing 95 percent of the cDNA molecules was probably harder than 95 percent of the genome since the two molecules varied wildly in abundance and, unlike the genome, cDNAs had far less evident measures of completeness. Craig responded "Thanks—I think." Around this time I was working for CRI as a consultant. I hatched a youthful prank to slip synthetic codes into the TIGR plasmid prep pipeline that might be decoded much later. Something similar seemed to plague a later effort to sequence DNA from pristine Atlantic Ocean samples, which instead found what looked more like Atlantic City, with genome messages in bottles resembling human sewage.

Follow Your Dream

A fanatic is a person who, upon losing sight of his goal, redoubles his efforts. This is the story of EngeneOS, Codon Devices, and Gen9. The dream was to do for biology what Intel had done for electronics. In 2001, Joseph Jacobson, Eric Lander, Daniel Wang, Stephen Benkovic, and I formed the founding scientific advisory board of EngeneOS (Engineered Genetic Operating System).

The company was one of the first to be financed by the Newcogen Group (later Flagship Ventures), a venture capital firm that had been the brainchild of Noubar Afeyan. Noubar had made some big money on an invention in the field of separation science, and in his time has founded or cofounded more than twenty life sciences and technology start-ups. One of them was Celera Genomics, where Craig Venter did some of his first work on sequencing the human genome.

The EngeneOS website claimed that its "technology platform starts with the 'source code' of Nature's operating system, embodied in the genomic sequences of various organisms. The company is combining this information with modern molecular biology techniques, engineering and design principles to develop Engineered Genomic Operating Systems. These systems will consist of component device modules supported by modeling and design tools."

In other words, EngeneOS expected to build a library of proprietary modular components, including engineered cells and proteins, as well as hybrid devices composed of biological and nonbiological materials. These modular elements would contribute to the design and fabrication of programmable biomolecular machines with novel form and function. It was an ambitious program: start with nature's operating system, reprogram it, and collect your output in the form of fabulous new engineered organisms.

So what happened?

Nothing. EngeneOS spun off various parts of itself before it ever really got started. The final spin-off was in 2004, when I, together with Drew Endy, Keasling, and others, cofounded Codon Devices, of Cambridge, Massachusetts, to develop tools for large-scale gene synthesis, CAD, and synthetic biology applications. Codon was very much a redo of EngeneOS. Our vision was not to compete with existing custom synthetic DNA firms, but to focus on the new opportunities in complex biological systems and miniaturization of processes, analogous to VLSI (very large-scale integration) in computer chips. We started with Samir Kaul as acting CEO. He was an inspiring leader in part because he had experience in the science of genomics and had worked with Craig Venter and his team on the first genome sequence of a plant (the mustard relative, *Arabidopsis*).We cornered the market on intellectual property related to DNA error correction as well as the thought leaders in the new field, the BioFab 9 mentioned earlier: David Baker (the undisputed king of protein prediction and design), myself, Jim Collins (genetic switches), Drew Endy (T7 refactoring, iGEM, the biobrick parts registry), Joe Jacobson, Jay Keasling (metabolic engineer), Paul Modrich (mismatch repair), Cristina Smolke (riboregulators), and Ron Weiss (genetic circuits for spatiotemporal patterning).

Unfortunately, despite all this intellectual horsepower, Codon Devices too fizzled and restructured. The CEO replacing Samir really liked the idea of a short-term payoff, as did several board members, and so they emphasized the idea of competing with existing companies that were doing DNA sequencing and synthesis rather than creating a market for something that didn't exist.

Later, Codon undercut the prices of other companies, plus they had a great sales force. But the sales force was so good at bringing in orders and so good at undercutting everybody else's prices that it started operating at a loss per customer. But even though the losses were very small per customer, they started adding up.

Codon Devices was disassembled in 2009. Nevertheless, we rescued parts of it in the form of yet a third start-up try at the original concept, this time called Gen9bio Inc. The idea was to start with DNA microarray chips on which gene-size (500 to 1,000 base pair) strands of DNA could be assembled. This sort of synthetic gene-building capacity would be used to produce both a large set of enzymes that are useful in making pharmaceuticals, and a set of constructs for optimizing overproduction of proteins in industrial-scale mammalian culture.

The company is so new that it just recently established its own web page. But it does have a few million dollars in initial financing and, of course, high hopes.

Follow Your Dream but Be Nimble

In 2001 Genomatica began as a metabolic engineering software company. Bernhard Palsson, Christophe Schilling, and others at UCSD had pursued a particularly useful brand of systems biology that combined tools from economic optimization with the detailed pathways of metabolism. The cell is like an industry, having a few choices of input materials, various ways to convert them into intermediate chemicals and, finally, end products. Given constraints on various transport and manufacturing speeds, one can adjust the entire network to maximize production of one particular product. In 2006 Genomatica morphed into a full-fledged synthetic biology company including wet lab experiments and a scale-up strategy.

The company aims to use its proprietary engineered *E. coli* bacteria to produce sustainable, green chemicals, at lower cost and with a smaller footprint than the products of conventional chemical companies. The initial chemical, Bio-BDO (1,4-butanediol), is used to make spandex, automotive plastics, running shoes, and insulation, among other things, and has a $4 billion market worldwide. The company has produced Bio-BDO at the pilot scale, and in 2011 was ramping up to demonstration-scale production. Genomatica has at least $84 million in financing and by all appearances looks to be one of the emerging leaders in the sustainable chemicals industry.

And now for a genuine and detailed synthetic genomics success story. In 2004 Jay Keasling, from the University of California–Berkeley, and one of the BioFab 9, received a grant from the Bill and Melinda Gates Foundation to make *E. coli* bacteria (and later baker's yeast) produce a precursor drug for the antimalaria drug artemisinin. Starting at zero, a few foreign genes not found in yeast had to be found and introduced, resulting in nonzero but still minuscule amounts of a useful intermediate product. The gene was resynthesized to achieve better codon usage, and indeed this improved yield 142-fold. Next, the yeast mevalonate pathway was optimized, resulting in a 90-fold greater output. Debugging yielded another 50-fold increase. Then optimization of the methods of fermentation (not the genome) gave another 25-fold boost. Finally a very clever idea of tethering some key enzymes via a scaffold normally used for very different purposes gave a shockingly high additional 75-fold improvement. Multiplying these factors to get to a billion-fold improvement in yield says more about how low we started than about how far we have come. For the latter, the better measure is how close to the theoretical maximum we are.

A general rule of thumb for maximum yields (as seen for Dupont's PDO process; see Prologue) is 100 grams per liter and 3 grams per liter per hour. Artemisinin precursors are typically made at 1 gram per liter, so theoretically there is some room for improvement here. In 2008 Jay's company, Amyris Inc., granted a royalty-free license to this technology to pharmaceutical giant Sanofi-Aventis for the manufacture and commercialization of artemisinin-based drugs, with a goal of having them on the market by 2012.

In part due to Jay's success in bringing this previously uneconomic drug to a cost point suitable for developing nations, in 2007 British Petroleum committed $500 million for synthetic biology research on biofuels at Berkeley. (Chapter 4 is devoted to the biofuel applications of genome engineering.)

In a Gold Rush, Sell Shovels

The million-fold plummeting costs of sequencing (Figure 7.1) could have meant the end of profit if customers hadn't thought of something to do with the sequences. One might call it a "shovel-rush." ABI had carefully groomed its monopoly for years and then a bunch of yahoos came in without a plan. Remarkably, the field responded by increasing demand by more than a million-fold, as if the auto industry started selling cars for $0.03 instead of $30,000 and people responded by ordering 2 million cars per household. And begging for more! For reading DNA, some of us had confidence (since 1977) that there would be a market of between 1 and 6 billion people with 6 billion base pairs each (over 10^{19} bp). The response to an analogous drop in costs of writing DNA is less well articulated—until now.

Let's zoom back to Figure 7.1 for synthesis. The synthesis of gene libraries and genomes begins with a source of short bits of DNA. Har Gobind Khorana created the first RNA oligomers in the early 1960s to help crack the genetic code (depicted in Figure 3.2), winning a share of the Nobel Prize in 1968 for this. Then he led his team to synthesize the first gene, which happened to encode the molecule at the core of the code and at the core of ancient and current life (our old friend tRNA). By the mid 1980s, Marvin Caruthers and team made a better chemistry that BioSearch and ABI automated so that labs could make one to four oligos just by typing in the sequence. In 1996 Blanchard, Kaiser, and Hood and later Rosetta Inpharmatics and Agilent adapted ink-jet printers to print A, C, G, and T onto flat glass slides. Xiaolian Gao at Xeotron and another group affiliated with Nimblegen came up with ways to make custom oligo arrays on the fly using spatially patterned light. Both groups teamed up with Jingdong Tian in my group in 2004 to show that the DNA on those arrays didn't have to stay on the arrays to be useful. Kosuri and coworkers published in 2010 a way to make subpools from the oligo arrays. And the freefall in cost of gene and genome synthesis suddenly seemed as inevitable as what had just happened for genome reading.

By 2012 the combined throughput of MycroArray, Combimatrix, LCscience, and Agilent could be around 300 billion base pairs per day, slightly behind the global genome sequencing capacity. Most of this is used for disposable arrays of oligos used for RNA quantitation or purification of genome subsets for sequencing—not for synthesis of genes or genomes. Clearly a market existed for sequencing 10^{19} genomes (a billion people with 6 billion base pairs each, plus repeat customers due to microbial, immune, and cancer genomics).

What might the markets be that will drive similar levels of consumption of DNA from chips? Here is our wish list (or bucket list, to go with the shovels):

Antibodies and fusion proteins
Binding proteins for DNA and RNA
Cell circuits: enhancer/splicing cis elements
DNA nano structures: smart drug delivery, nuclear magnetic resonance rods
Enzymes: every type and metagenomic
Foreign DNA: de novo or ancient reconstructions
Genomes: new codes, new amino acids, virus resistance
Homologous recombination: integrases
Isolation and safety chassis
Joined sensor-select: +/-allosteric regulators
Knowledge, media, data storage, steganography
Ligand engineering
Metagenomic access
Nanopore sequencing and sensors
Opto-electronics and scaffolding
etc. . . .

The point is that the applications of large-scale DNA writing are even more up for grabs than DNA reading. Predicting the future development of this field would be like trying to guess what applications of the personal computer would have been in the early 1970s: Electronic recipe books? Ping-pong? Balancing your checkbook?

ۏ   ۏ   ۏ

Synthetic biology is mostly about developing and applying basic engineering principles—the practical matters that help transform something academic, ivory-towerish, pure, and sometimes self-indulgent or abstract into something that has an impact on society and possibly even transforms it. Systems biology moves us from massive numbers of observations to theories. But when you try to build something—even an academic something—you really acid-test those ideas—often finding out how little you need to understand or how much you do need to understand but don't. When you try to build something for society, it's even tougher since many things that work in your lab don't work in other labs, much less in the hands of regular folks. And even if they do work, there is no guarantee that they will be financially successful.

The business of synthetic biology may now be transitioning to making a living by synthesizing genomes. What has been largely missing is an articulation of why we should engineer whole genomes rather than just the important parts. In answer, I have described a project to change the genetic translation code genome-wide for safety, to create new amino acids, and to engineer resistance to all viruses.

We are already in the business of making a safer microbial "chassis." In 2011, DARPA put out a request for proposals for ways to "watermark" pathogens being actively studied in laboratories so that we could more easily trace accidental or intentional releases. This reflects our 2001 DARPA proposal to use DNA as a storage medium, and as we will see in Chapter 8, embedding English, encrypted messages, or even images in DNA has a history going back at least to 1984. The new challenge is to make these messages stable across time. Another DARPA challenge is to make the pathogens being studied able to survive only under specific laboratory conditions and not in the wild—without at the same time altering them so drastically that researchers can't study their pathogenicity. All of these measures also apply to nonpathogenic, synthetic organisms, that although considered safe, would likely be more widely (and wildly) distributed (because of industrial utility) and hence more worthy of possessing tracking and safety features.

In Chapter 2 we examined ancient human texts and compared their longevity to the texts of life. In the tradition of encoding art in DNA discussed above, the world's first so-called synthetic organism (Craig Venter's M. mycoides) was accompanied by bits of human text embedded in code (the four letters of DNA). One bit of text read: "To live, to err, to fall, to triumph, to recreate life out of life." This sentence, which is a quote from *A Portrait of the Artist as a Young Man*, by James Joyce, prompted the Joyce estate to send Venter a cease-and-desist letter. This beautifully captured the moment at so many levels. Was quoting the Joyce text "to err, to fall" (i.e., was it an error on Venter's part?) or, to the contrary, was including the text in a historic hunk of DNA "to live, to triumph" (i.e., to glorify Joyce)?

In addition to the unfortunate Joyce quote, the genome of *M. mycoides* JCVIsyn1.0 also incorporated a misquote of what Richard Feynman wrote on his last blackboard. As Venter's team had it, "What I cannot build, I cannot understand." They were quoting from a secondary source (a risky business), because what Feynman actually wrote was: "What I cannot create, I do not understand." That too elicited a corrective note from the authorities, in this case Caltech, where Feynman taught, together with a picture of the blackboard in question. (The JCVI researchers further compounded their errors by expressing the Feynman word in using a code that allowed only uppercase letters, which netizens interpret as shouting. But FEYNMAN WAS NOT SHOUTING! Nor, for that matter, did he write his parting message in caps.) Finally, we note that "What we can create, we don't necessarily understand."

I noted that small blooper when I assessed the JCVI manuscript for *Science* prior to its publication in that journal. To lighten up the rest of my critique, I playfully submitted my review to *Science* encrypted entirely in DNA. I heard later that Clyde Hutchinson, an investigator at JCVI, was the one who figured it out. For the first few sentences, I used their code, but then I switched to a code that allows the encoding of anything digital, including lowercase letters, images, audio, and even web pages—and is easier to recall than sixty-four codons. In alphabetical order A = 00, C = 01, G = 10, T = 11. Even if the code is easy to recall, will the encoded messages endure, perdure, or neither?

The English poets Percy Bysshe Shelley and Horace Smith told the poetic tale of Pharaoh Ramesses II (c. 1303 to 1213 BCE), whose fate has much in common with the fleeting glory of coded quotes and misquotes. Here is Shelley's version:

I met a traveler from an antique land
Who said: "Two vast and trunkless legs of stone
Stand in the desert. Near them, on the sand,
Half sunk, a shattered visage lies, whose frown,
And wrinkled lip, and sneer of cold command,
Tell that its sculptor well those passions read
Which yet survive, stamped on these lifeless things,
The hand that mocked them and the heart that fed.
And on the pedestal these words appear:
'My name is Ozymandias, King of Kings:
Look on my works, ye Mighty, and despair!'
Nothing beside remains. Round the decay
Of that colossal wreck, boundless and bare
The lone and level sands stretch far away."

Will our DNA watermarks flow and evaporate with time? Or, worse yet, what if we leave an indelible, hubris-laden watermark graffito on our planet that says only "To err"?

CHAPTER 8

-100 YR, ANTHROPOCENE
The Third Industrial Revolution. iGEM

In early 2010, three undergraduate biology students at the Citadel—Brian Burnley, Patrick Sullivan, and Hunter Matthews—had an idea for a swell genomic engineering project: they wanted to reprogram the *E. coli* bacterium, which is a normal and benign resident of the human gastrointestinal tract, so that it would secrete a peptide that would suppress appetite. Obesity, they knew, was one of the top health problems in the United States, as well as in a lot of other countries. The problem affected one-third of the American adult population, and increasingly it also affected children. To combat it, people tried all sorts of weight reduction schemes, bogus and genuine: they took metabolism-revving pills, they dieted, they exercised, they bought stationary bicycles, ellipticals, and other exercise-torture machines, and they had excess fat removed from their bodies by liposuction. But in many cases nothing worked—or at least not permanently.

Why not tackle the problem where it arose, at the source, inside the human intestinal tract? *E. coli* bacteria, which can be made to do practically anything, were already on the scene. That's where they lived. If they could be genetically engineered to express satiety peptides, which controlled hunger, then the whole obesity problem could be addressed biologically. Indeed, almost magically: the microbes would control your appetite for you, effortlessly, with no conscious intervention on your part, much less any heroic dieting willpower.

It sounded like a panacea, almost too good to be true. This would be a new type of cure-all, with programmed bacteria working on the problem inside your body. Obviously the idea would not appeal to everybody: people who were already on the warpath over the issue of genetically modified foods would never go for the idea of genetically modified bacteria rummaging around in their intestines. Still, the scheme provided a novel way for people to take control over their own physique.

All that was needed for the project to succeed were the genes necessary to persuade *E. coli* to synthesize PYY (3–36), a peptide naturally produced by mammalian colon cells and that is responsible for satiety regulation. Dr. David M. Donnell, an assistant professor of biology at the Citadel, knew where to get the necessary genetic structures: from the Registry of Standard Biological Parts at MIT. So toward the end of June he ordered them. And about two weeks later they arrived in the form of freeze-dried powders in 384-well plastic plates covered by adhesive foil seals. A list accompanying the shipment identified the exact well that each desired biological part was in. To use a given part, the experimenter inserted a pipette through the foil covering the well in question and injected into it a small amount of liquid. The powder containing the part dissolved readily and the solution was then ready to be drawn out and moved to a centrifuge for further processing.

So Donnell and the three students, aided and abetted by Dr. Claudia Rocha, also of the Citadel's Biology Department, proceeded with hands-on microbial genetic tinkering. If and when they got the scheme to work—and even if they didn't—they would enter their project in the 2010 iGEM competition.

iGEM stands for international genetically engineered machines. The name had a stirring, thrilling, almost chilling ring to it. It was reminiscent of *R.U.R.*, a science fiction play by the Czech author Karel Capek. It premiered in 1921, and gave the word "robot" to the world, for R.U.R. stood for "Rossum's Universal Robots." Capek's robots were not the metallic, tin can, electronics-driven devices of today. Instead, they were biological machines that had been assembled in a factory, part by part. In the play, unfortunately, the robots take over the world and wipe out the human race—a tiresome and overworked scenario if there ever was one.

iGEM, by contrast, was a real-world organization in which student teams from around the world competed in building novel biological machines, also part by part. These "machines" were in fact microbes, but they were so substantially altered and enhanced by the genes of other organisms that they constituted new and original types of microorganisms, which performed specific, predictable tasks not normally executed by natural biological systems.

For example, these engineered organisms could be made to detect arsenic levels in drinking water; they could be formed into thin films on which visual patterns and images could be deposited, after the manner of a photographic film; they could produce hemoglobin that could be used to replace red blood cells in emergency transfusions. Or, just possibly, they could be fashioned into automated weight-control devices.

These new microbes were machines in more than a merely metaphorical sense. For one thing, they were built out of standardized parts, prefabricated genetic building blocks that could be reassembled in almost any desired order or combination. For another, these parts were to be put together according to a uniform or standard protocol for connecting one component to the next. In consequence, if the resulting combination of prefabricated, standardized biological parts worked at all, it worked predictably and reliably, just like a machine. That, anyway, was the idea.

To bring the idea of standardization into genomic engineering—the standardization of genetic parts, assembly methods, and combination procedures of the type advocated in the BioFab Manifesto—would change the way we viewed, understood, and used biological systems. Originally, you pretty much had to take organisms as they came, with all the inherent design flaws and limitations, compromises and complications, that resulted from the random workings of evolution. Now we could actually preplan living systems, design them, construct them according to our wishes, and expect them to operate as intended—just as if they were in fact machines, *appliances*.

These ideas—organisms as machines, part by part construction, standardization—have their counterparts in the third industrial revolution, which is the one we normally think of when we use the term "industrial revolution."

৯৯   ৯৯   ৯৯

The third industrial revolution (1750–1850) was one of the great turning points in human history. Key elements of the transformation were the change from artisanal, custom-made, hand-tooled methods of producing material goods to machine-tooled, assembly-line, and standardized mass production systems. These changes allowed for unprecedented levels of income growth and wealth accumulation, sustained increases in agricultural production, human population growth, and enhanced well-being.

The third industrial revolution was a product of at least two separate technological developments. The first was the use of steam to power machinery, and this goes back to ancient times. It began over the period from 250 BCE to 50 CE with Ctesibius (pronounced *teh-sib'-ee-us*), and then Hero (his real name) and his "aeolipile" device (*its* real name). Ctesibius, an inventor and engineer living in ancient Alexandria, was a maestro of air and water. He built an air-powered catapult, a water-powered pipe organ, and a water clock that worked by dripping water into a container at a constant rate: a float with a pointer attached to it marked the passage of time on a vertical scale. He also described a device for pumping water out of wells.

But it was Hero of Alexandria who invented one of the first steam engines. He attached two bent tubes to a hollow sphere, filled it with water, and heated it over a fire. The tubes then became steam nozzles and caused the sphere to spin (this was his aeolipile). Although the device did no practical work, it was nevertheless an apparatus that converted steam into motion, and was thus a true machine—a steam turbine. (Among other things, Hero also invented the world's first vending machine, a Rube Goldberg apparatus in which a set amount of holy water would be dispensed after a coin was deposited through a slot at the top.)

Practical use of steam awaited Englishmen Thomas Savery in 1698 and Thomas Newcomen in 1712. Savery's steam engine had no piston or other moving parts and worked instead by generating steam inside a vessel and letting it condense by cooling, which induced a partial vacuum that drew water up through a tube.

Newcomen constructed a mechanical engine in which a giant arm was made to reciprocate back and forth by filling a chamber with steam at about ten times atmospheric pressure. This drove a piston up inside the chamber, raising one end of the arm until, at the top of the stroke, the pressure was released through a valve. The process was then repeated. These devices were typically used for pumping water out of mines and were widely employed and widely imitated. In 1775 James Watt, the Scottish inventor and mechanical engineer, attempted to replace water as the obligatory source of power for mills. This freed the mills from the need to be located near streams.

The other development that made the industrial revolution possible was standardization. Earlier, when individual craftsmen turned out their wares, each item was a product of the artisan's own personalized set of hand tools—chisels, files, scrapers, planes, hammers, saws, and so on—all of them used according that individual's particular level of skill, working methods, and often idiosyncratic design of the finished product. The result was a vast range of made-to-order manufactured objects, each of which was mechanically incommensurable with all the others. Even the very nuts and bolts, the screws and nails, that held the object together differed sufficiently from one clock maker or gunsmith to the next that in many cases one artisan could not repair an artifact made by another. This system was not a model of optimization.

But on April 21, 1864, William Sellers, a toolmaker who was also president of the Franklin Institute in Philadelphia, read before that society a paper that would become one of the landmark documents of the machine age. Titled "On a Uniform System of Screw Threads," it proposed a system in which every nut, bolt, and screw would be of uniform length and diameter, and the threads would be angled at a precise and consistent 60-degree pitch. Such threads were easy to measure and manufacture (60 degrees being the angle between the sides of equilateral triangles), and they made for an exceptionally strong and secure fit. This uniformity of size and thread made for a system of fully interchangeable parts.

The Sellers system soon became a benchmark, then a national standard, and finally an international standard. It was a paradigm case of sensitive dependence on initial conditions, for his seemingly fussy, minor, even trifling innovation was the crucial design change that allowed the hand tool age to give way to the machine tool era characterized by interchangeable parts, reliable operation, and assembly-line mass production.

In twenty-first-century America, the William Sellers of biology was an MIT computer science whiz, artificial intelligence engineer, and information omnivore by the name of Tom Knight. He arrived at CSAIL (MIT's Computer Science and Artificial Intelligence Laboratory) as a fourteen-year-old high school student one summer in the 1960s and apprenticed himself to legendary artificial intelligence sage Marvin Minsky. Knight would go on to invent, coinvent, and play supporting roles in developing a number of the iconic structures of the computer age, including hardware for ARPANET, LISP machines, and Danny Hillis's Connection Machine.

Later, in the 1990s, Knight came across the work of Harold Morowitz, a Yale physicist and biologist. (A man of considerable dry wit, Morowitz was the author of several popular science books, including *The Thermodynamics of Pizza* and *Mayonnaise and the Origin of Life*.) Morowitz had a special interest in a category of bacteria called (seriously) *Mollicutes*, the most generally known of which were the *Mycoplasmas*, which were also the smallest and simplest known free-living, self-replicating life forms. They consisted of a countable number of atoms (about a billion). Morowitz thought, and so did Knight, that studying these ultra-simple organisms would reveal the underlying architecture of life.

"Morowitz's work laid out, in words I could understand as an engineer, an agenda that seemed so exciting that I had to go do it myself," Knight once explained. "Here's a class of organisms so simple that maybe we can understand everything there is to know about them."

Being very much a hands-on, let's-find-out type of guy, Knight set up a biology lab inside of a little niche within CSAIL where he could do experiments on *Mesoplasma*, a species closely related to *Mycoplasma*. He quickly discovered that organisms, even small and simple ones, did not operate with the predictable crisp precision of computer electronics or well-written software. And there was nothing routine or standardized about the process of experimenting on them.

Every time he tried to make a DNA construct for an experiment, he said, "it was done in a different way, depending on what plasmids were available, what restriction sites were present in the fragments being assembled, what cell lines were available for transformation. The list goes on and on, and the more you know, the longer the list is. It drove me crazy, as it would any self-respecting engineer. The path forward was clearly to standardize the part definition and the assembly process."

So, in essence, that is what he did. In a paper titled "Idempotent Vector Design for Standard Assembly of Biobricks" (2003), Knight proposed the idea of a standard biological part, which he called a biobrick, as well as a standard procedure for putting two or more of them together. Each biobrick would be a circular piece of double-stranded DNA (otherwise known as a plasmid) containing a smaller sequence of DNA base pairs of known structure and function. The smaller sequence was the biological part of interest, whereas the rest of the plasmid was a sort of backbone or framework. Each end of the smaller sequence (the "prefix" and "suffix") consisted of a specific molecular structure that made for a sticky end that could join with the corresponding sticky end of any other biological part.

The standard procedure for assembling two biological parts was simple enough conceptually (Figure 8.1). Using the restriction enzymes and the cutting and joining techniques commonly employed in molecular biology, you would cut biological part **A** out of the plasmid that contained it, and slice open a space at the sticky-end site of the plasmid containing the second biological part, **B**. Then you would mix together part **A** and the sliced-open plasmid that contained part **B**, thus allowing the respective sticky ends of parts **A** and **B** to join up. This created an expanded plasmid containing the new biological part **AB**. That plasmid could then

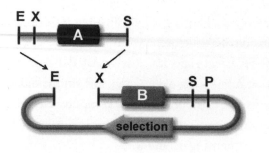

Figure 8.1 Biobrick assembly

be put into *E. coli* cells, which would express whatever it was that the new part coded for.

The assembly process, Knight commented in his paper, was such that "each reaction leaves the key structural elements of the component the same [which is what the "idempotent" of his title meant]. The output of any such transformation, therefore, is a component which can be used as the input to any subsequent manipulation. It need never be constructed again—it can be added to the permanent library of previously assembled components, and used as a compound structure in more complex assemblies." This was standardization as implemented biologically.

❧   ❧   ❧

Knight's idea of creating a "permanent library of previously assembled components" was the basis for the Registry of Standard Biological Parts, an inventory of snap-together, prefabricated genetic Lego blocks. Although Knight founded the library, much of its later development was substantially the work of Drew Endy.

When it was formally set up in 2003, Endy was a fellow in the Department of Biological Engineering at MIT. Like Knight, Endy did not start out in biology, and in fact had a love-hate relationship with the subject. He didn't much care for either the inherent complexity of natural organisms or their native tendencies to sporadically come forth with unexpected, "emergent" properties and behaviors.

"I hate emergent properties," he has said. "I like simplicity. I don't want the plane I take tomorrow to have some emergent property while it's flying."

What Endy liked about biology was that organisms were so versatile and pliable that you could coax them to build practically anything simply by manipulating their genetics.

"I like to build stuff," he says, "and biology is the best technology we have for making stuff—trees, people, computing devices, food, chemicals, you name it."

Unsurprisingly, Endy became a biological engineer, getting his PhD in biochemical engineering and biotechnology at Dartmouth. In 2003 Endy, who had won a teaching award at Lehigh in 1993, offered an independent activities period (IAP) course at MIT. Such courses were routinely given on offbeat subjects during the January break, and Endy's was definitely in the mold: Synthetic Biology Lab: Engineered Genetic Blinkers. Over the previous eight months, Endy had developed the course materials together with Tom Knight, Gerald (Gerry) Sussman of the AI Lab, and Randy Rettberg, who was a research affiliate of the AI Lab but had previously held executive positions at Sun Microsystems and Apple Computer.

The course, which was attended by sixteen students ranging from freshmen to doctoral candidates, was aimed at producing a genetic sequence that would make *E. coli* bacteria periodically emit a dim luminescence, flashing on and off like a stoplight (albeit one that flashed in exceptionally slow motion). Making bacteria fluoresce was nothing new in microbiology: genes for the green fluorescent protein (GFP) normally found in jellyfish were routinely spliced into an organism's genome for the purpose of creating a "reporter gene," which signaled that a given molecular event had occurred within the organism. A blinking bacterium was nothing new, either. Michael Elowitz and Stan Leibler, then at Princeton, had created a three-part circular genetic loop in *E. coli* that made the bacterium oscillate regularly between "on" and "off" states. The gene for the on state also activated a green fluorescent protein gene that made the organism light up, a bacterium that blinked on and off. The controlling element of the object was a combination of oscillator and repressor genes, and so they called their new genetic circuit a "repressillator." (Their report on the device was published in *Nature* in 2000: "A Synthetic Oscillatory Network of Transcriptional Regulators.")

So, what was new about Endy's IAP-engineered genetic blinkers project in 2003? Basically, the idea was to turn the original genetic blinker into a new and improved version with superior functionalities. The oscillatory period of the Elowitz and Leibler blinker was measured in hours, which was longer than the cell division cycle of the bacterium itself. Consequently the state of the oscillator had to be transmitted down from generation to generation. So one subgroup of IAP students wanted to speed things up. A second subgroup wanted to create and install a "synchronator" gene that would make all the cells blink off and on in concert, like the lights on a movie marquee.

The students started with an initial biobrick parts kit put together by the instructors, Endy, Knight, Sussman, and Rettberg. The parts kit consisted of actual DNA sequences stored in freezers, sequences that encoded some of the most basic functions in molecular biology. There were promoters: sequences that started DNA transcription, and there were terminators: sequences that stopped transcription. There were antisense RNAs that blocked gene expression, and of course there were GFP genes that made cells fluoresce. A data book listed the function, structure, device name, description, and biobrick number pertaining to each separate genetic part. Essentially, the parts kit was an indexed bin of biological components and supplies.

Standard biological parts exist on three levels of complexity; the most basic is a genetic part such as a plasmid or a plasmid backbone. There are also composite parts, combinations of two or more individual parts, such as a reporter+quencher. The next level up is a genetic device, a group of components that perform a function of greater complexity, such as a device that initiates cell death or sends signals between cells. Finally, there are systems, entire organisms that do something—blink on and off, for example. (There are also "chassis" parts, which are commonly used lab strains of organisms such as *E. coli*, yeast, or *Bacillus subtilis*. These are more or less blank slate microbes that will receive the new genetic sequences and then express them as output.)

During the IAP course, the students came up with and contributed about fifty new genetic parts of their own, and entered each part's name, function, and other information in the data book. This was the basis for

MIT's current Registry of Standard Biological Parts, which can be accessed at parts.mit.edu. At the time the IAP course ended, there were about 140 annotated genetic parts in the registry. At the time of this writing there are over 7,000 parts in the registry, but this is changing quickly, since Harvard's team alone made 55,000 parts in the summer of 2011. Even the location and governance of the registry is changing, moving away from the MIT campus.

When the students were finished designing their new genetic circuitry, Endy sent the relevant sequence information to Blue Heron Biotechnology of Bothell, Washington, to be synthesized. The company, founded by MIT grad John Mulligan in 1999, found that half of the designs couldn't be produced, and those that could be synthesized didn't seem to work when injected into cells. And so the whole project seemed to be a washout.

But about a year later, Endy managed to get some of the designs to function as intended—just barely. "It would have been dumb luck if they all worked together out of the box," he said.

The other instructors had their own postmortems. "It's interesting to think about where this is going, and of course we have no idea," said Gerald Sussman. But Tom Knight did have one: "Hopefully, the next IAP students will get *E. coli* to blink 'dash-dash, dot-dot, dash' (Morse code for 'MIT')."

The course was repeated in January 2004, with a project named Engineered Genetic Polka Dots. Student teams created new standard biological parts and designed genetic circuits that made cellular patterns including polka dots (the team in question called itself the Polkadorks) and bull's-eyes, as well as an ambitious plan to make cells swim together like a school of fish.

In the end, none of these second-generation grand designs behaved as planned, either. Evidently, putting biobricks together and getting them to work successfully would be slightly more challenging than anyone had thought.

ꙮ   ꙮ   ꙮ

But then came iGEM.

In the beginning, iGEM stood for intercollegiate genetically engineered machines. The idea was for MIT to bring other American colleges into the business of engineering genetic parts and devices, and to stage a competition among the participants. The competitions would be modeled on student robot contests, events that were so successful at drawing in students that the prize ceremonies at some of them filled entire stadiums. So Rettberg, Endy, and Knight rounded up Caltech, Boston University, Princeton, and the University of Texas–Austin (where Endy had spent some time as a postdoc) and delivered them all a one-phrase challenge: design and build a genetically encoded, finite-state machine (one that transitions from one state to another under the control of a program).

The students would start work in the summer of 2004, continue through the fall, and then meet up for a jamboree in Cambridge early in November to compare and share their results. Which is indeed what happened.

The Princeton team set out to build the biological equivalent of a children's game called Simon, a test of memory in which participants had to repeat a pattern of signals that grew in length and complexity with each successive iteration. So the students designed a three-cell system that, when it detected a particular sequence of inputs, would respond by triggering a release of the reporting molecule, which in this case was yellow fluorescent protein (YFP).

The Caltech team wanted to design and build a strain of yeast that could detect three different levels of caffeine in coffee. Boston University teamed up with Harvard to design cells that could be made to count. The UT Austin contingent set out to build what sounded like the neatest bacterial app of them all, an *E. coli* film that would take what amounted to a biological photograph.

As it turned out at the jamboree, the Princeton team couldn't get its YFP reporting mechanism to work. The Caltech students, by contrast, brought off their project handily. Their engineered yeast glowed green in the presence of low caffeine, green and yellow in medium caffeine, and yellow alone in high caffeine. In a campus coffee shop, the yeast successfully identified decaf, regular, and espresso coffees. The Boston University and Harvard team eventually got their bacterial counter to work and published a report on the project (together with considerable subsequent work) in *Science* five years later.

The stars of the show were the Texans, who added photoreceptor genes to *E. coli*, created a thin film of the light-sensitive bacteria, and got it to display the phrase "Hello World" in the form of deposited pigments (Figure 8.2). This phrase has been a standard simple test for computer programmers and an inside joke ever since Brian Kernighan wrote a Bell Labs memo in 1974. The experimental result was published in *Nature*: "Synthetic Biology: Engineering *Escherichia coli* to See Light" (2005). A

Figure 8.2
Hello World

behind-the-scenes story is that the original pigment patterns were due to the action of the light directly on the colored nutrients in the agar. In other words, the photo was a product of the nutrients rather than the bacteria. The team had not done the control step of leaving out the genetically engineered microbes. Nevertheless, that photo was all over posters representing iGEM for the next year! But this lapse was repaired before the paper was submitted to *Nature*. (Team members also sharpened the fonts and added the word "nature" as well as faces of many almost-famous people, like Andy Ellington of Austin, Texas.)

Undergrads were now doing things, largely in a spirit of fun, that professional molecular biologists would have been hard-pressed to achieve a mere ten years earlier.

The following year, iGEM attracted teams from Canada, England, and Switzerland, and participation has grown steadily since then. The 2006 iGEM finally took on the trappings of a formal competition, with judging panels, prizes, and categories such as best part, best device, best system, best presentation, and even "best conquest of adversity." The jamborees themselves, meanwhile, increasingly resembled Olympic-level sporting events, with competitors showing up in team costumes, carrying away trophies, and building human pyramids on the playing field to celebrate victory.

Some of the genetic systems they designed worked so well that they had commercial possibilities. Students from the University of Edinburgh won the 2006 iGEM prize for the best real-world application with a bacterial arsenic sensor for drinking water. Whereas existing assays had a sensitivity of 50 parts per billion (ppb), the team's *E. coli* sensor could detect concentrations of arsenic as low as 5 ppb. In the 2007 competition, the team from UC Berkeley engineered *E. coli* to produce a blood substitute that could be freeze-dried and stored, and then could be reconstituted and grown up in large volumes when needed. In 2008 the grand prize winner and official recipient of the Biobrick Trophy, a large silver Lego block bearing the iGEM logo on one face, was the team from Slovenia (which had also been the winner in 2006), which this time created a synthetic vaccine for the pathogen *Helicobacter pylori*, the cause of stomach ulcers.

❦　❦　❦

The 2010 iGEM Jamboree turned out to be the biggest and best of them all—the last one in which all the teams met and competed at a single location, which as it happened was iGEM's birthplace and continuing home base, MIT. As a venue for a synthetic biology intercollegiate competition, MIT was spiritually, emotionally, and symbolically correct: it was internationally renowned as one of the world's top educational and research institutions for techies, computer geeks, and all-around science, technology, and engineering fanatics. During World War II, MIT's Radiation Laboratory ("Rad Lab") was home to pioneering work on inventions such as radar and LORAN (long-range navigation). More recently, it has been noted for work in robotics, machine learning, cybernetics, and the theory and practice of artificial intelligence.

The Rad Lab connection brings to mind the fourth industrial revolution—the electro-magnetic industrial revolution. This was precipitated in 1870, when James Clerk Maxwell built models that unified electricity, magnetism, and light. He is best known for stating Maxwell's equations—a set of four equations, each of which, somewhat comically, was stated by and named after others. Maxwell's equations are, in the usual sequence, Gauss's law for electricity, Gauss's law for magnetism, Faraday's law of induction, and Ampere's law with Maxwell's correction. These equations have a quirky visual appeal to them, and are popular on T-shirts, with a top caption reading "And God said," followed by the equations and a subcaption: "and then there was light." There's also a poster saying: "What part of" (the equations) "didn't you understand?"

Maxwell's equations model the relationships between electricity and magnetism as well as the wavelike behavior that constitutes electromagnetic radiation. "Then there was light" includes not only visible light but also radio waves, microwaves, infrared, ultraviolet, and cosmic rays. The huge success of this unification of two fields of physics set the stage for later successes in unifying the main forces in physics and aiming us toward grand unification theories. Oh, and a very prolific set of civilization-changing inventions, like television, satellites, and cell phones.

It's hard to overstate the significance of electromagnetic radiation in the ancient evolution of life, ranging from the development of photosynthesis to the genesis of vision, and in the future of nanobiotechnology, running the gamut from photolithography to the creation of optical sensor networks.

However ideologically suited MIT was to an iGEM jamboree, physically the place was a maze, a vast, sprawling assemblage of industrial and anonymous-looking buildings that run for about a mile along the Cambridge side of the Charles River. It takes a while to get spatially oriented to the layout, because at MIT one building looks pretty much like the next, and indeed in many cases several of the buildings physically merge with one another to form a seamless megastructure. As a further aid to confusion, MIT's buildings tend to be known by number rather than by name (though each does have a name). The total effect on a newcomer is akin to being dropped into a secret government intelligence-gathering complex located in a hitherto undiscovered foreign country. In his book about MIT, *Up the Infinite Corridor: MIT and the Technical Imagination*, Fred Hapgood observes that the practice of referring to buildings by number fosters the impression that "they were just larger rooms in a single enormous building. . . . Building 7 feeds into 3 and 3 sits next to 10 and 10 next to 4, and so on. A stranger rushing to make a scheduled appointment might think the design calculated to drive him crazy. . . . Any point in the campus seems equally near or far from any other."

This impression of being lost in space was relieved for some of the 2010 iGEMites by the fact that many of them had been there for previous jamborees and thus they could make sense of the geographical master plan. The tournament started on Friday, November 5, 2010, when 130 iGEM teams, comprising about 1,300 students, arrived on campus for the three-day event. They came from all reaches of the globe and included thirty-eight teams from Asia, ten from Canada, thirty-eight from Europe, thirty-seven from the United States, four from Latin America, and one from Africa. (No longer was iGEM confined to college and university students: the 2010 competition featured a team from Gaston Day School, a prep school in Gastonia, North Carolina. Its project, "Construction of a Biological Iron Detector in a Secondary School Environment," was to build an engineered organism that could detect high levels of iron in water sources.)

The teams brought an array of biological engineering schemes, plans, and molecular designs that ranged from the whimsical to the incredible, and from the trivial to serious attempts to address global medical problems by means of artfully rewired microbes. A team from the University of Bristol was pursuing "smarter farming through bacterial soil fertility sensors." A collaborative effort between Davidson College in North Carolina and Missouri Western State University put microorganisms to work on solving the "knapsack problem." (Given a set of weighted items and a knapsack of fixed capacity, is there some subset of these items that fills the knapsack?) A group from the Swiss Federal Institute of Technology at Lausanne aimed "to stop malaria propagation by acting on the vector: the mosquito . . . We are engineering *Asaia*, a bacterium that naturally lives in the mosquito's gut, to express an immunotoxin that can prevent the malaria agent, *Plasmodium falciparum*, from infecting the mosquito, thereby eliminating transmission of this parasite to humans."

The iGEM team from Polytechnic University of Valencia, Spain, had a plan to change the climate of the planet Mars. "We are going to build an engineered yeast resistant to temperature changes and able to produce a dark pigment which will be responsible for a global temperature increase on Mars." (How's that for redesigning nature?) The team from the University of Washington–Seattle was bent on synthesizing a range of novel antibiotics for the twenty-first century. "Using synthetic biology tools," they reported, "we designed, built, and tested two new systems to fight infections . . . Our first project targets *Bacillus anthracis*, the Gram-positive pathogen that causes anthrax."

The competition got under way at 10:00 AM Saturday, a gray day with a cold wind blowing in from the river. Because the teams were so numerous, and because each team was allotted a full thirty minutes for its presentation, the talks could not be given in a single linear stream, sequentially. Instead, they were split into six divisions that ran concurrently in six different MIT buildings. This meant that six separate groups of judges would gather at the end of the day to compare results and establish rankings.

An iGEM team presentation consisted of a number of essential elements, one of the most important being that the proceeding started and ended exactly on time. The next crucial element was the team uniform, which typically took the form of a brightly colored T-shirt bearing a logo, a bacterial design, and in some cases corporate endorsement patches such as might be worn by an Indianapolis 500 racecar driver. PowerPoint slides, and/or videos, were of course an integral part of things. The final necessary element was frequent use of the word "so" to start a sentence, whether or not the sentence actually followed a previous one or had any kind of logical, organic, or conceptual relation to it. (This is a verbal tic common to geeks of all types, stripes, flavors, and ages.)

A canonical presentation was offered by the Chinese University of Hong Kong, whose team consisted of ten students, each of whom took a turn in giving a segment of the talk, which was held in 26–100 (Building 26, Room 100). The team's goal was to convert *E. coli* bacteria into information storage devices, something on the order of microbial flash drives— or as they called them, bio-hard disks. The group titled its abstract "Bio-cryptography: Information En/Decryption and Storage in *E. cryptor*," *E. cryptor* being the team's name for its designer *E. coli* microbe, which members also referred to as "a living data storage system." They presented a scheme by which all 8,074 characters of the US Declaration of Independence could be encoded, encrypted, and stored in engineered *E. coli*, and then decrypted, decoded, and retrieved back as text.

The first part of their talk was given over to their techniques for translating characters of plain, written text into quaternary (base four) numbers, and then mapping each of the quaternary numbers onto one of the four chemical bases of DNA. For example, the letter H in plain text would correspond to the number 1020 in the quaternary number system, which in turn would be expressed as TACA in DNA (where 1 is represented by T, 0 by A, and 2 by C). Using such correspondence principles, the plain text word "hello" would become 10201211123012301233 in quaternary encoding and TACATCTTTCGATCGATCGG in DNA encoding. (A program for converting any given text message into quaternary numbers and then into DNA nucleotides can be found at 2010.igem.org/Team:Hong _Kong-CUHK/Model.)

The students described additional techniques for data compression, for deleting repeated sequences, and for ensuring the accurate representation of the message by means of a checksum algorithm. Using these and other means, the team members had calculated that it would take eighteen individual bacterial cells to reliably store the full Declaration of Independence in *E. cryptor*.

Their offering ended with the conjecture that, since essentially any type of information can be digitized, it will one day be possible to reliably store not only text but also pictures, music, and even video—in bacteria. Even among iGEM projects, which are characterized by bold, out-of-the-box thinking, this scheme was notably ambitious.

There were some precedents for writing human language text and images into DNA, however. In 2009 Claes Gustafsson wrote a paper for *Nature*, "For Anyone Who Ever Said There's No Such Thing as a Poetic Gene." In it, he described how his company, DNA2.0, Inc., during the Christmas season of 2005 gave away free synthetic DNA that encoded the first verse of "Tomten," a poem by Viktor Rydberg. The verse amounted to fifty words, about eight hundred base pairs long. The protein sequence was back-translated to DNA using the codon bias of reindeer (*Rangifertarandus*; Well, it was Christmas!). Gustafsson's final claim was: "To our knowledge, this is the first example of an organism that 'recites' poetry."

An even earlier precedent was set by Joe Davis in 1984–1988 (described in a 1996 article), who cloned a 28-mer length of DNA representing a 5x7 pixel line drawing. W. Wayt Gibbs in his 2001 *Scientific American* piece, "Art as a Form of Life," also described the process and gave additional insight into Joe's work as well as various earlier Nobel-level molecular engineering pranks.

Going forward, we could amplify libraries of 200 mers by carefully minimizing variation in abundance using flanking universal primers. The design would have extensive adjacent sequence overlaps. This can be made much easier to read than a shotgun human genome, for example, by using (1) precise overlaps instead of ragged, random overlaps and (2) extra base pairs to disambiguate repeats. (3) Compression algorithms (as used for Internet images) can also be used to reduce the number of repeats. (4) Check-bits and other tricks mitigate synthesis and PCR errors. The primers at the ends are designed to enable immediate plug-and-play compatibility with standard next-generation sequencing. We will skip two of the hard steps—assembly of oligos at the beginning and the fragmentation and library-making at the end. High synthetic redundancy and similar levels of analytic (sequencing) redundancy help ensure low error rates.

Some hints on what to digitize in an intentionally representative earth message can be seen from the 200 megabyte golden record on the Voyager spacecraft that was launched in 1977. We can store millions of copies of data sealed in small (optionally phosphorescent) plastic time capsules around the world. Nearly every major company, person, and copyright holder would want to be represented in these time capsules.

So, putting pictures, music, and even video into bacteria, as the Hong Kong iGEMites suggested, is not out of the question. As Donald Trump once said, "If you're going to be thinking at all, you might as well think big." In that spirit, Wikipedia all-languages version amounts to 53 GBytes, the DNA version of which would be 90 billion base pairs long. At 1 bp per cubic nm this would fit into a 5 micron diameter sphere (the size of a human red blood cell) at 100X redundancy and would cost about $1 per 10^5 copies. By comparison, Blu-ray video disk digital storage is on a 12 cm diameter CD that is 1.2 mm thick, and contains 50 GBytes (which is fully a billionfold bulkier than DNA per bit). At $1 per disk in bulk, the disks would be 100,000-fold more expensive than DNA. Compared to what you can do with biology, Blu-ray data storage is a real waste of money and space! Believe it or not!

✿   ✿   ✿

Presentations continued throughout the next day, Sunday, starting at 9:30 AM and ending at 5:00 PM. That night, there was an enormous party (the iGEM 2010 Jamboree Social Event), held at Jillian's, an epic food, drink, billiards, bowling, sports, dancing, and entertainment paradise adjacent to Fenway Park, located across the Charles River in Boston. (Free shuttle bus service was provided to and from the saturnalia.)

The awards ceremony was held the next day, Monday, starting at 9:00 AM, in the Kresge Auditorium of Building 46, which was a curvaceous, glassed-in structure designed by Eero Saarinen and resembled what a hangar for flying saucers probably ought to look like. The auditorium seated 1,226 people, approximately equal to the total number of students, faculty advisers, visitors, photographers, videographers, and other members of the media who were in attendance.

Randy Rettberg, the director of iGEM, was master of ceremonies. After the ritual return of the BioBrick Trophy from the prior year's Grand Prize winner (the University of Cambridge), the judges announced the six finalists whose names were projected on a screen: the University of Bristol, Cambridge, Imperial College London, Peking University, Slovenia, and the University of Technology, Delft.

Clearly Slovenia was the team to beat. Based at the University of Ljubljana, Team Slovenia had been the grand prize winner at two previous iGEM competitions in 2006 and 2008, with impressive projects in each case: in 2006, a scheme for preventing infection of human cells by HIV, and in 2008, the development of a synthetic vaccine for *Helicobacter pylori*.

The six finalists now had to repeat their presentations before all the competitors and all of the judges. This took nearly three hours, under hot lights, general restlessness, and rising ambient temperatures. After a break, the assembled iGEMites, dazzling in their team colors, gathered together for the rip-roaring "iGEM from Above" group photo (2010.igem.org /Main_Page).

The results were announced in reverse order of importance, accompanied by the distribution of Best Application and other "Best" prizes, prizes for the runners-up, and so on, until, finally and at last, the grand prize winner was revealed to be . . . Team Slovenia.

And appropriately so, for their project this year recapitulated, on the level of DNA engineering, the way in which assembly-line mass production had improved upon the prior system of individual craftsmen working by themselves. Whereas craftsmen turned out their products slowly and laboriously, the assembly line had organized the production of a given object into an orderly, fast, and efficient process that furthermore occurred in a single place, such as in a factory.

Similarly, DNA expression normally took place as a result of enzymes randomly bumping into and around the molecule and then fitting in when and where possible. This was wasteful and inefficient, and could be improved. And so the Slovenians vowed to make DNA engineering more like an assembly-line process. First the team members created a custom DNA molecule by putting specific sequence blocks together in a predefined order. Then it enhanced a group of enzymes by means of proteins that enabled the enzymes to bind selectively to given, predefined sites on the molecule. The custom DNA and the newly enhanced enzymes significantly increased the speed and yield of the DNA expression reaction. Finally, by changing the order of the DNA sequence blocks, the team could change the reaction's output. The result, they announced, was "DNA guided assembly lines."

Essentially, it was the industrial revolution all over again, but played out this time on the level of molecules. In the course of doing all this, Team Slovenia submitted 151 new biobricks to the registry.

Team Citadel, as it turned out, did not figure in the rankings. The three team members, Brian Burnley, Patrick Sullivan, and Hunter Matthews, had taken their turns, giving an abbreviated presentation on Sunday, the last day of the event, in the last time slot of the day. Their talk was abbreviated because, owing to delays in getting their design synthesized, they hadn't completed the construction of their appetite-suppressing bacterium. The students learned the hard way that advertised DNA synthesis times were decidedly on the optimistic side, for in their case what was supposed to take three days instead took three weeks. In consequence, their obesity-controlling bacterial devices—which they had since named Appetuners—did not meet all of the criteria required for formal recognition at the jamboree.

The Chinese University of Hong Kong, by contrast, received a Gold Medal for its "living data storage system." And so did Team Valencia for its scheme to terraform Mars with strains of dark-colored yeast.

Team Harvard also received a Gold Medal for its iGarden project. Its aim was to enable gardeners, either with or without a scientific background, to genetically engineer their own plants—to "personalize" them according to their own individual tastes and preferences. Team members had taken a poll of regular folk gardeners and found that the most commonly requested plant improvement was to increase the nutritional value of fruits and vegetables. This could be done, for example, by enhancing the gene for lycopene or beta carotene production.

To prevent an engineered plant from spreading beyond its intended boundaries, the Harvard team members advanced the idea of building a molecular "fence" around an iGarden. This would take the form of a "death gene" that would be activated if and when the engineered plant escaped from the lab. They hoped that this strategy would help allay some people's negative attitudes toward genetically modified foods. (Of course, it might also cause them to run screaming from their greenhouses.)

With the end of 2010 competition, iGEM reached a turning point. It had become so popular, and participation had become so large and unwieldy, that the jamboree could no longer be held all at one venue. IGEM 2011, therefore, would be split into three regional divisions, Asia, Europe, and the Americas, with separate competitions and regional jamborees held within each division. The winning teams would then advance to the World Championship Jamboree at MIT. (The grand prize winner at the 2011 iGEM was the University of Washington, whose team project had the dual goals of engineering *E. coli* to produce diesel fuel and engineering an enzyme to break down gluten in the digestive tract.)

The growth of the iGEM competition surpassed the wildest hopes of its creators. What it all meant was that, increasingly, some of the world's most imaginative, significant, and potentially even the most powerful biological structures and devices were now coming not from biotech firms or from giant pharmaceutical companies, but from the ranks of university, college, and even secondary school students, who were doing it mainly in the spirit of advanced educational recreation. Proof of the power and allure of redesigning life.

CHAPTER 9

-1 YR, HOLOCENE

*From Personal Genomes to
Immortal Human Components*

❦

The Holocene epoch has lasted from about 10,000 years ago to the present. During this period human civilization as we know it arose and blossomed. One of the most important elements of civilization, across all cultures, times, and regions, has been the development of medicine, a system of beliefs intended to explain and influence the course of the main stages and changes that characterize human life: birth, death, health, and disease. Back in the most ancient eras, these events and conditions were often regarded as the work of spirits, demons, gods, and various astral or mystical phenomena. (Belief in faith healing is still with us today.)

The history of medical folkways is unsavory, gory, and even gruesome, filled with wild beliefs, horrifying practices and devices, and useless nostrums and potions, all thought to promote health or cure illness. Bloodlettings, libations, purgations, emetics, piercings, shamans, the ritual sacrifice of animals or humans, concoctions featuring eye of newt and/or other ground-up animal parts, various and sundry snake oils, brews, spices, tinctures, tonics and elixirs, syrups, salves, rubs, pills, drops, lotions, airs, diets, spells, incantations, rites, sacred dances, trances, chants, sexual practices, special bodily ornaments and charms, tricks with magnets, and the like—all of these and more are treasured elements of medical (mal)practice through the ages.

Western medicine, as a science (as opposed to folklore), is commonly viewed as starting with Hippocrates, who was associated with a medical school, the Asklepieion, on the Greek isle of Kos. Unlike those who attributed disease to occult entities or forces, Hippocrates offered physical, rational, and empirically founded explanations of health and disease, which he furthermore viewed as physical states or conditions. The subsequent history of medical science is the story of reducing health, disease, and bodily functions to their natural and material underpinnings. Well-known medical milestones include Vesalius's description of human anatomy, William Harvey's theory of blood circulation, and Pasteur's germ theory of disease. But as important as they are, all of these advances are dwarfed by the advent of molecular genetics, and in particular by the decoding of the human genome.

The human genome is the recipe for building and maintaining a human being, and its component genetic structures, discrete sequences of DNA, are the underlying molecular sources of disease and health. More than three thousand human diseases are known to be caused by a specific gene or a combination of them (acting in concert with the environment), and nowadays hardly a week goes by without an announcement of the discovery of a new gene for a given disease or condition, everything from asthma to zoophobia.

To speak of a "gene for a disease" is actually to speak of a *mutation* of a gene that would otherwise not play a role in causing illness. The gene for cystic fibrosis, for example, was discovered in 1989 by a team headed by Francis Collins (whom we have already met as head of the Human Genome Project). In cystic fibrosis, cysts and fibrous scars in the pancreas block the pancreatic ducts, restricting the flow of digestive enzymes. The intestines and lungs can also be involved, in the latter case resulting in respiratory infections. Collins and his team located a long DNA sequence on chromosome 7, and the sequence later became known as the *CFTR* gene.

It normally codes for a protein that transports salt and water across the cell membranes of various organs. The Collins group found that the deletion of just three bases, CTT, from the normal DNA sequence of the *CFTR* gene was sufficient to cause the disease. (Other researchers later discovered more than 1,000 other mutations that can also play a role in causing cystic fibrosis.)

Other illnesses are caused by the difference of as little as a single base in an otherwise normal genetic structure. Sickle-cell anemia, the classic example, is the result of the substitution of just one letter of the HBB gene sequence that codes for normal hemoglobin. The HBB gene for normal hemoglobin includes the triplet GAG, which is the three-letter code for glutamic acid. In mutant hemoglobin, however, the triplet GTG, which codes for the amino acid valine, appears where GAG should be. The occurrence of the base T (thymine) instead of the proper base A (adenine), a difference of a single molecular structure on the gene, is enough to induce the malformation of the blood cells characteristic of sickle-cell anemia. For want of a nail good health is lost, here in the form of a severe, chronic, and generally incurable condition. (The variation of a single base from what's normal is known as a single nucleotide polymorphism, or SNP.)

As these examples show, we can now pinpoint, with literally atomic accuracy, the molecular basis of many human pathologies. For many such cases, we are reaching a major goal, in the reduction of health and disease states to their ultimate physical foundations in the human genome. The discovery of "dark matter" in the genome, material distinct from known genes that plays a causal role in some sicknesses, is of ongoing interest.

Thus the importance of the Human Genome Project as well as that of its latter-day descendant, the Personal Genome Project.

&&   &&   &&

Our ability to trace states of disease and health to their atomic underpinnings is a manifestation of the fifth industrial revolution, which focused on atoms. This revolution originated in the discovery and applications of strange quantum phenomena in physics and chemistry. In a single year (1905) a twenty-six-year-old patent clerk, unable to get a job in academia, published four papers (all in the same journal, *Annalen der Physik*). Any one of these papers could have earned him a Nobel Prize, and indeed the first one did do so. The topics were the photoelectric effect (in which light photons dislodge electrons, the outer components of atoms, from a solid or liquid surface), Brownian motion (in which small particles are buffeted by the thermal vibration of molecules), special relativity (which pertains to motion at nearly maximal speeds), and the mathematical relation between mass and energy. This last paper included what is without a doubt the most well-known equation of all time, $E = mc^2$. It is famous both because of its brevity and for its Promethean recipe for the release of earth-shattering amounts of energy resulting from the splitting or fusion of small concentrations of atomic nuclei. Understanding these phenomena required measurements of matter interacting with light (atomic spectra) and other kinds of particle beam radiation.

The quantum revolution has made an enormous impact on chemistry and materials sciences. These effects can be roughly divided into nuclear and electronic phenomena. The radius of an atomic nucleus is about 100,000 times smaller than the radius of the surrounding electron cloud that defines most of the atom's chemical properties. Quantum-mechanical breakthroughs concerning electronic bonding have been fundamental to understanding the nature and strength of chemical reactions and interactions. Exploration of the nuclear realm enabled the introduction of radioisotopes, which have been crucial for many advances in biochemistry and medicine as well as the dating of ancient specimens, and in DNA sequencing. The use of stable isotopes and mass spectrometry extended such studies considerably. The diffraction of X rays from material objects ranging from simple salts to organelles like the ribosome, as well as information from the interactions of nuclear spins in nearby atoms in NMR, have provided a window into the world at atomic resolution, paving the way for precise molecular engineering.

The damaging biological effects of ionizing radiation were discovered by the American biologist Herman Joseph Mueller in 1926. Work done by Mueller and others in the early days of studying X rays, as well as natural and artificial radioactivity, has helped us understand mutations that occur spontaneously due to cosmic radiation from our sun and many other celestial sources dating back to the dawn of time. These mutations occur despite the protective effects of the atmosphere (equivalent to about 10 meters of water) and the Van Allen belts, two regions of charged particles that partly surround the earth at heights of several thousand kilometers.

The unwelcome consequences of the atomic revolution include nuclear war and radiation sickness, the Chernobyl meltdown, and the possibility of worldwide nuclear holocaust. Negative aspects of chemistry include pollution, drug abuse, and the problems posed by semiconductors (a subject covered in my discussion of the sixth revolution), computer viruses, identity theft, privacy invasion, cyberwar, and bioterror.

❧   ❧   ❧

The original Human Genome Project sequenced the 3 billion base pairs of the human genome at a cost of $3 billion, or at an average cost of one dollar per base pair. That was a milestone of science, but its significance was offset by two factors: its high cost and the fact that the genome that had actually been sequenced was not the DNA of any one individual but a composite genome of many DNA contributors. It was the sequence of a "blended" person, and so it had little value in practical, personal, or medical terms. It was the moon landing of molecular biology. (The genomes of two individual humans differ by an average of about 3 million positions, which is approximately 0.1 percent of the total. Most of these are single base changes or changes in tandem repeat lengths.)

Although the genomes of any two people are 99.9 percent identical, the genetic differences between them account for much of their physical uniqueness, including predisposition to illness and differential responses to drugs, medical treatments, and infectious disease agents, as well as their psychological individuality and personal tastes, preferences, talents and deficiencies.

Hitherto, medicine has operated largely on a one-size-fits-all approach, tailoring a cure to the disease rather than to the person who is suffering from it. For a long time this made sense: a disease, after all, is a specific thing and human beings are genetically almost carbon copies of each other, and so what alleviates a disease for one person ought to perform equally well for the next. But often enough it doesn't. A drug that helps one person may be toxic to another, may provoke an allergic reaction or have other adverse side effects, or may have no effect whatsoever. Such differential responses are often found with respect to antidepressant medications, for example, many of which can take two weeks or more to have an effect, if any. Discovering a genetic basis for these varying outcomes would allow doctors to prescribe drugs that worked most effectively for a given person. Indeed, such discoveries are now laying the groundwork for the new field of pharmacogenomics.

People also respond differently to the same disease agent. The bacterium *Staphylococcus aureus*, for instance, kills an average of 100,000 Americans per year, more than any other single microorganism. It is the leading cause of heart, skin, and soft tissue infections, and is a common cause of pneumonia. It is the top causal agent of nosocomial (hospital-acquired) infections. Nevertheless, some 30 percent of the population harbor the pathogen in their nasal passages but show no sign of infection. Evidently there are genetic factors at work that explain these dissimilar responses to the microbe. This finding has implications for the future of medicine. If a patient's genome sequence were part of his or her electronic medical record, and susceptibility to *Staph* infection was known in advance, then the subject could be treated with appropriate antibiotics before being admitted to a hospital where the infection might otherwise be acquired.

The original Human Genome Project was made possible by the then emerging niche technology of automated DNA sequencing machines. Ten years after the success of the HGP, improvements in that technology have brought down costs to levels at which commercial personal genomic services have become a reality, and one day a complete human genome sequence will be available for about $1,000. This will inaugurate the era of new approaches to health and disease, an era of personalized genomic medicine.

As a teenager, I had the grand notion that we ought to sequence every-body—all 6 billion base pairs for all the 4 to 7 billion of us—and store the data in computers. This was a sort of "genomes for all" approach, to be pursued for predictive reasons alone. The idea was that if you knew the types of diseases or medical conditions you were predisposed to—adult-onset diabetes, let's say—then you could take appropriate countermeasures early in life. Given the cost of both sequencing technology and computers in those days, that plan was naive, to say the least.

But today, when the cost of computers and automated DNA sequencing technology continues to plummet, my plan is not so naive. Whereas the data storage capacity and processing speeds of computers has tended to follow Moore's Law over the past fifty years, with the number of transistors on integrated circuits doubling every year and a half, the cost-effectiveness of DNA sequencing has increased by about ten times per year over the last six years.* Such improvements are only likely to accelerate, and con-sequently the sequencing of whole populations at low cost soon will be possible.

I still think that, ideally, all who desire it should be able to have their genome sequenced, and for predictive reasons first and foremost. After all, there are already about two thousand known, actionable, and highly pre-dictive genetic associations. Even though they may be rare, they are never-theless predictable and actionable—conditions that you can do something about.

Another reason to sequence everybody is to create a database that shows correlations between genotype and phenotype—between a person's genome and the set of observable characteristics that result from the in-teraction of the person's genome with the environment. (As geneticists like to say, the genes may load the gun but the environment pulls the trigger.)

---

* Here I use the conventional formulation of Moore's Law. In 2000 Gordon Moore stated: "I never said eighteen months. I said one year, and then two years . . . Moore's Law has been the name given to everything that changes exponentially in the industry. If Gore invented the Internet, I invented the exponential."

There are correlations not only between genes and diseases but between genes and observable traits such as eye color, hair color, facial features, cognitive abilities, eating habits, lifestyle, personal history and experiences, career choices, mental outlook, and lots of other things. The genomic database would be an immense toolbox for understanding the myriad ways in which genes and the environment interact to form the sum total of human individuality and variability.

This dataset would have to be fully open to researchers, to any investigator who wanted to use it for any purpose whatsoever, whether to generate hypotheses, run tests, establish or disprove correlations on any level, or anything else. That would mean open publication of the individual's genome and phenotype on the Internet, available for all the world to see. In effect it would be putting your life story, medical history, and genetic makeup on the web. It would be the Facebook of DNA.

ℬ   ℬ   ℬ

This brings us to the sixth industrial revolution—the information-genomics revolution. Like the others, this revolution has crucial quantitative measures—probability, information, and complexity—as well as possibly crucial emerging measures of life, evolution, and intelligence. Theories of probability began as ruminations on how to win at games of chance, as illustrated by the *Liber de ludoaleae* (Book on Games of Chance), written by Gerolamo Cardano in 1526. Information and communication theory builds on probability theory. The basic concepts were first clearly articulated by Claude Shannon in his paper "A Mathematical Theory of Communication" (1948). One of Shannon's goals was to quantify the amount of information lost to static and other influences in phone line signals. He called this measure of uncertainty "information entropy" as an analogue to entropy in thermodynamics, where it refers to the irreplaceable loss of heat energy.

John von Neumann, the mathematician and sometime wit, liked Shannon's term, telling him: "You should call it entropy, for two reasons. In the first place your uncertainty function has been used in statistical mechanics under that name, so it already has a name. In the second place, and more important, nobody knows what entropy really is, so in a debate you will always have the advantage." (This was a concept with legs, as the idea was later exported to the business world as "corporate entropy"—energy lost to bureaucracy and red tape.)

This sixth revolution is allowing us to understand, generalize, and make connections to the previous ones. For example, the precise sequences of DNA, RNA, and proteins discussed in Chapter 1 have a great deal in common with the strings of zeros and ones of the digital computing revolution.

Some unwelcome consequences of the sixth revolution include computer viruses, identity theft, privacy invasion, cyberwar, and bioterror. The potential scale of cyberwar became evident recently in the extent to which the Iranian computers controlling isotope separations could be compromised by individuals having no physical access to the devices (possibly Israeli sympathizers). The cost to society of hacking, Trojan horses, and computer viruses and worms is about $50 billion per year. Bioterrorism is currently small to nonexistent, but the stakes are astronomically high.

ᔥ   ᔥ   ᔥ

The original goal of the Personal Genome Project (PGP) was to sequence the genomes of 100,000 volunteers at no cost to them, and to publish the results on the Internet along with each individual's personal data, even down to their picture. Of course, any such plan immediately raises privacy issues. The PGP couldn't promise to keep any of this information private, since the whole point of the exercise was to create an open access source of genetic and personal information and to disseminate it as widely as possible.

The solution was to accept into the program only people who, like myself, consider that the benefits to society outweigh the risks—and also regard privacy as a highly overrated asset. But then the question was, How could we guarantee that we'd recruit only such people? The answer was to formulate a list of eligibility criteria, create consent forms, and devise an online entrance exam composed of a few dozen questions that would measure each volunteer's understanding and acceptance of the totally free and open-access nature of the enterprise. A potential recruit would have to give correct answers to each and every question, the equivalent of getting a score of 100 percent on an examination. The candidate could take the test as many times as necessary to get 100 percent, but a perfect score would be required for enrollment in the program.

In addition to passing the entrance exam, participants would be required to sign two consent forms: a five-page mini consent form outlining the program and its requirements and constituting a basic eligibility screening. A second and more elaborate sixteen-page full consent form would describe the program, the candidate's participation, and public release of the data generated, in great and heroic detail.

All this was acceptable to Harvard's Institutional Review Board (IRB), the body responsible for approving, monitoring, and reviewing medical research or experimentation on humans. In August 2005, the IRB gave us approval for a pilot program. I became the first candidate.

Today, anyone can go to PersonalGenomes.org and view my public profile, which includes vital signs (my height, weight, and blood pressure), allergies (none), medications (lovastatin, coenzyme Q, multivitamins, calcium, etc.), medical history (narcolepsy and squamous cell carcinoma, among other fun things), race (white), traits (male, blood type O+, green eyes, etc.), facial photographs (suitable for framing), DNA data sets, and type of tissue samples taken (lymphoblasts and fibroblasts), plus date collected, storage location, and accession number. All of this information is followed, furthermore, by a universal waiver, stating in part: "To the extent possible under law, PersonalGenomes.org has waived all copyright and related or neighboring rights to Personal Genome Project Participant Genetic and Trait Dataset."

My tissue samples were taken in 2005 and 2006. My lab has developed or advised most of the current thirty-six commercial next-generation sequencing technologies, and we test these technologies as they mature. The first set of samples was sequenced at Complete Genomics in Mountain View, California.

The results were underwhelming. My genome should have shown alleles for narcolepsy, dyslexia, high cholesterol, cardiac arrhythmia (and maybe musical arrhythmia as well!), squamous cell carcinoma, and plantar fasciitis. But it didn't (yet). Sequencing a genome is one thing, but interpreting and understanding it—making sense out of the practically endless and visually meaningless strings of the nucleotide letters A, T, C, and G—is quite another. Doing so requires the use of software that translates those otherwise baffling chains of letters into usable, practical information. The process of developing such software is still in its early stages. In December 2010 the University of California–Berkeley hosted a competition for genome interpretation programs. It was called the inaugural Critical Assessment of Genome Interpretation (CAGI) competition and attracted more than one hundred entrants. The very existence of this competition shows that we have a long way to go before a genome sequenced is a genome understood.

ꝗ   ꝗ   ꝗ

The Personal Genome Project was formally opened to the general public on DNA Day of 2009, April 25, the anniversary (recognized by the U.S. Congress) of the day in 1953 that Watson and Crick's paper describing DNA's structure was published in *Nature*. The first ten participants, myself and nine others, became known as the PGP-10. The Harvard IRB initially wanted the first ten candidates to have at least a master's degree in genetics, but the board later dropped this requirement as impractical for subsequent scaling up. In the end, the PGP-10 included Esther Dyson (PGP-3), who describes herself as "a longtime catalyst of start-ups in information technology," Steven Pinker (PGP-6), a Harvard psychologist, and Misha Angrist (PGP-4), an assistant professor at Duke, the only one of the PGP-10 who has a PhD in genetics.

In 2009 Pinker wrote a first-person account of his participation in the project for the *New York Times Magazine*, "My Genome, My Self." The piece described a decision that every potential PGP participant had to face: whether or not to learn that you may be carrying the gene for an incurable disease such as Huntington's or Alzheimer's. Pinker chose not to learn whether he had a variant of the *APOE* gene that would predispose him to Alzheimer's disease. (This is known as "redacting" the information in question.)

He did learn that he had one copy of a gene for familial dysautonomia, an incurable disorder of the nervous system with unpleasant consequences, including premature death. "A well-meaning colleague tried to console me," Pinker wrote, "but I was pleased to gain the knowledge."

His other genomic discoveries included good news and bad news. The good news was that he had a less than average chance of getting prostate cancer before age 80. The bad news was that he had a slightly elevated chance of developing type 2 diabetes. (He had a risk of baldness, despite the fact that he had great hair.)

Pinker's results show that genetics is not always destiny. Some genes are deterministic. If you have the gene for Huntington's disease and you live long enough, you will sooner or later develop it. Otherwise, the influence of genes on traits is probabilistic, stacking the odds in one direction or another but without completely predetermining the outcome. Better yet, since there is a strong environmental component of our genetic destiny, we can take action to influence or defeat it.

Another of the PGP-10 also wrote about the journey through his personal genomic universe. This was Misha Angrist (PGP-4), who published the story in his book, *Here Is a Human Being: At the Dawn of Personal Genomics* (2010). His data had been interpreted by the Trait-o-matic, open-source software developed at the Church lab for the purpose of finding and classifying the ways in which genetic sequence variations manifest themselves in the human body.

When Angrist logged on to the website holding this personal genomic information, he too discovered a modest number of unimpressive genetic data points about himself. Those data points, however, are now part of a growing body of openly accessible set of genotypic-phenotypic correlations, a sort of medical analogue to the World Wide Web.

There is a difference worth noting between the Personal Genome Project and the group of private companies that do over-the-counter genome analysis for the masses. The genomic information provided by the Personal Genome Project is useful both to the individual sequenced as well as to the wider community of researchers, with the dual goals of transforming the practice and delivery of medicine and understanding how genomes give rise to living beings and how they influence the manifold processes of life.

Commercial genome sequencing firms, like Knome, by contrast, hold their data privately, and for good reason, for the information they acquire could be put to a number of unpleasant uses. For example, it could be used to infer paternity, affect employment or insurability, or even one's love life. Ironically, the information often is not used by the person for whom it was intended. A study conducted by Scripps Health, of La Jolla, California, and published in the *New England Journal of Medicine* in 2011, reported that out of 2,037 people whose genomes were analyzed by the private firm Navigenics, most of them had failed to make any changes in their diet or exercise patterns when they were interviewed three months after receiving their test results, even when their results showed a definite need for making such changes. (Still, 27 percent of the participants who shared their results with their physicians did make some lifestyle changes.)

In the end, just as individuals respond differently to drugs and to pathogens, they also respond differently to information.

ॐ   ॐ   ॐ

A volunteer for the Personal Genome Project donates saliva, blood, and a bit of skin. This allows checking for accuracy and for somatic mutations, epigenetic changes, and microbial components. ("Epigenetic" refers to the ways in which influences outside of strict DNA nucleotide sequences— for example, environmental factors—can modify gene expression.) The skin samples provide fibroblast cells that could be reprogrammed into synthetic yet personalized pluripotent stem cells that could be further reprogrammed into a variety of cell types—probably into all natural body cell types as well as many novel cell types and tissues of medical value. Their value includes personal tests for drug toxicity and efficacy, testing for inherited disorders in advance of their appearance, and generating perfectly histocompatible tissues for organ transplants without the need for immunosuppressant drugs. It makes gene therapy possible without the use of viral vectors (which caused cancer in some previous gene therapy trials). Two tipping-point events that the world barely noticed while they happened in 2007 (but were covered by Pulitzer Prize–winning articles recently) were the stem cell treatments of little Nic Volker and Timothy Ray Brown.

Nic Volker was a child living in Madison, Wisconsin, who had started having gastrointestinal problems just before his second birthday. Physicians at Children's Hospital in Milwaukee failed to come up with a definitive diagnosis, but nonetheless performed more than one hundred intestinal surgeries before Nic was four years old, without ever solving the problem.

"Normally with medicine we can get these problems under control," said Dr. Alan Mayer, a gastroenterologist. "But with Nic we never really did. And the disease continued to progress. It was so severe, my intuition told me this had to be due to a genetic mutation."

Mayer enlisted the help of genomics experts Howard Jacob and Elizabeth Worthey and had Nic's genome sequenced. Within four months, Jacob identified the gene that was responsible for Nic's condition, a rare disease for which a bone marrow transplant was the treatment of choice. This was done in July 2010, using stem cells from the cord blood of a matched, healthy donor, and the boy was cured.

"So what we really did was we replaced his immune system that was defective with a different immune system that lacked that defect," Mayer explained. Nic is fine today, a testament to the power of genomic medicine.

Forty-two-year-old Timothy Ray Brown lived in Berlin, Germany, and suffered from both leukemia and AIDS. Gero Hütter, a hematologist at Berlin's Charité Hospital, found a blood stem cell donor who was not only matched for Brown's tissue type but also had a rare double deletion of the CCR5 gene, which happens to be the host cell receptor for the HIV-1 virus. As a result of that transplant Brown has been free of both leukemia and AIDS (and without anti-HIV drugs) for the last four years. In a sense, he was the world's first person to be cured of AIDS.

Nic Volker's case was a turning point because now many parents concerned about developmental delays or other medical mysteries in their child will insist that their children be given the kind of genomic examination that Nic got. Brown's case is significant because his genome therapy made use of a rare genotype (the double deletion of the CCR5 gene, which is carried by only about 1 percent of humans). In principle, many of us might want to have this rare genome, just as we might want a vaccine to prevent AIDS, even if we aren't planning on engaging in risky sexual behavior. Sangamo BioSciences, a California biopharmaceutical company, has in the works a clinical trial of a method to change a patient's own stem cell genomes precisely and efficiently to inactivate both copies of CCR5. The results of the trial are promising so far. These stem cell transplants might some day be extended to include resistance to all viruses (see Chapter 5).

The pace of personal genomics' impact on health is accelerating as this book goes to print. The Noah and Alexis Beery twins misdiagnosis of cerebral palsy nearly thirteen years ago was corrected when their genome sequence revealed a mutated gene related to brain serotonin and dopamine production and fixed by dietary supplements (like 5-hydroxy tryptophan). The drug Ivacaftor (aka Kalydeco) was approved in three months by the FDA, a speed record we all hope will become typical. It is specific for a single base change in the gene responsible for cystic fibrosis. In the face of steady progress, we still hear echoes of cautionary notes that genetics diagnostics will generally be uninformative, but need to also note that most (nongenetic) diagnostics, like pulse rate and blood chemistry at your annual medical exam, will also not tell us anything new. But often enough these tests do have high predictability and actionability, so we all should check. Genetics seems to be similar.

ᆶ   ᆶ   ᆶ

The prospect of improved health through personal genomics raises the question of what the limits are to such improvements, and whether it makes sense to envision exceptionally long-lived human beings and even potentially immortal human components. Thoughts of immortality go back to the early Neolithic era, when humans began to think deeply about death, mortality, and everlasting life. The ancient heroes—pharaohs and kings—took every means available to ensure their own vestigial future persistence through the ages. They had themselves enshrined in legends, songs, poems, constellation names; they had their remains preserved in vast pyramids, in larger-than-life statues, and so on. Why?

Part of the reason may lie in the fact that we live so long already, on average. Our species is distinctive in our ability to remember and to predict future events based on past experience. (Granted, some animal species also do this, at least on a primitive level.) Before the invention of writing (and to a large extent even afterward), reliable prediction-making required the embodiment of these memories in a living person. People well past their reproductive years could add value to their tribe by remembering what to do when a rare phenomenon approached—a drought, locusts, war, and so on. In modern times, our training is even more extended, including postdoctoral or on-the-job training well into our sixties (or even longer). Unlike computers whose full memories can be moved to a new model, human memories and skills are much harder to transfer. Most modern diseases are caused by aging. Many researchers even argue that, with nations becoming wealthier and exposure to infectious agents and toxins dropping, aging will become the underlying cause of most diseases. Yet many people remain quite active beyond the age of one hundred. We have much to learn from these natural human long-lived specimens, and this is a clear opportunity for synthetic biology since the cure is probably not a drug but rather a redo of our genome. We can search the best of the biosphere for ideas.

As we have seen, species run the gamut when it comes to longevity. Mayflies live up to their insect order name. Among *Ephemeroptera* (Greek for "ephemeral wing") the adults live, dance, and mate for thirty minutes (after having lived for months as larvae). Gastrotrichs, a type of microscopic aquatic invertebrate, live out their entire life cycle in three days.

At the other extreme is the bowhead whale (*Balaena mysticetus*; see Chapter 3), which grows to 150 tons, second only to the blue whale, by feeding in the fertile but chilly Arctic. Estimates of individual bowhead whale longevity have ranged up to 210 years, judging from the age of ancient spear heads lodged in their flesh as well as the rate of mirror-flipping of amino acids in their eye lenses. The oldest fish of all, a koi, was a female scarlet specimen named Hanako (born c. 1751), which died at the age of 226 years on a memorable date: 7/7/77. Two radiated tortoises have set the land record: one named Tu Malila lived to 188 years (until 1965), and another, Adwaita, lived for possibly 256 years (dying in 2006). The ocean bivalve mollusk, *Arctica islandica* (also known as the quahog, or the hard-shell clam), can live up to 405 years, albeit in nearly freezing water where the rate of chemical changes, metabolism, and hence ensuing oxidative damage is very low. (As is, for that matter, the roster of hard-shell clam accomplishments.)

Trees, which are even more limited in their mobility and accomplishments than mollusks, can nevertheless live to the grand old age of 4,860 to 5,000 years. Indeed, the world record for oldest nonclonal organism is held by a bristlecone pine named Prometheus, which lived for literally centuries near Wheeler Peak in eastern Nevada. It was killed in 1964 by grad student Donald Currey, who was unfortunately bent on determining the tree's age by counting tree rings. By Currey's count, Prometheus was at least 4,862 years old when felled, and probably older than 5,000 years. (Wordsworth's famous line "We murder to dissect" was never so apt!)

But it is possible to go beyond old age. At least one researcher, Daniel Martinez, reported in 1998 that the hydra is one of the few animals that does not undergo senescence at all and is therefore biologically immortal. (In 2010 another researcher, Preston Estep, disputed Martinez's claim.)

Astonishingly, *Turritopsis nutricula*, a jellyfish, can actually *get younger*. It has the fountain of youth gift of returning from its sexually mature (medusa) state back to a younger (polyp) state. An entire population of such organisms can do this synchronously and swiftly—although it is difficult to observe in the wild. Still, in virtue of its unique ability to return to an earlier stage of life, *Turritopsis nutricula* have managed to escape biological death through aging (although members of the species succumb to predation, accident, and disease).

What this bizarre menagerie of extremely long-lived, possibly immortal organisms is leading up to is the fact that, arguably, human cells (if not people) can be immortal too. This is suggested by the case of Henrietta Lacks, an African American woman who died of cervical cancer on October 4, 1951, at Johns Hopkins Hospital in Baltimore, at the age of 31. For cancer research purposes cell samples had been taken from her cervix. They were code-named HeLa cells, after the first two letters of her first and last names.

Then the cells took on a life of their own. Hopkins researcher George Gey found that HeLa cells could be kept alive and grown in lab glassware. The cells replicated, grew, and proliferated so wildly and uncontrollably that they often took over and wiped out any other cell lines they happened to come into contact with. Indeed, HeLa cells are still alive today, all around the world, many decades after they were removed from the cancerous tumor of Henrietta Lacks.

On the other hand, HeLa cells are so biologically aberrant that they have seventy to eighty-six chromosomes (which works out to an average of 82, amid the midst of chaos of chromosome breakage, joining, deletion, and duplication), rather than the usual forty-six, in twenty-three pairs (which in contrast to the indeterminacy of HeLa cell chromosomes is an exquisitely precise number). Furthermore, HeLa cells are far from being able to produce normal human tissues.

The human cells that are both immortal and capable of making humans immortal, or at least longer-lived, are germ cells. Some large multicellular creatures (many plants, for example) do not have an opposition between germ line cells and somatic cells. But for many animals, including humans, the germ line is the only part of us that naturally survives us in our offspring. (Germ line cells are the body's reproductive cells, or gametes; the rest of the body's cells are somatic cells.) Indeed, germ line cells are the all-time champions of cellular survival. We can trace their DNA back through billions, possibly trillions of binary divisions. In 99.999+ percent of those past divisions, only one of the cellular siblings survived, or if both survived then eventually all of the offspring of one died and only our ancestor survived the bloodbath.

Is cloning a possible route to immortality? Over twenty species of animal have been cloned, including carp, mice, sheep, rhesus monkeys, gaurs (wild oxen), cattle, cats, dogs, rats, mules, horses, water buffalo, and camels. Some of these clonings are routine procedures with a clear agricultural benefit. Cloning is sometimes viewed as dangerous, but many new technologies go through a phase in which they are actually unsafe, or at least perceived as unsafe, and the technology is banned at least locally. This is often followed by a phase in which the public demands the technology. For example, railroads. (Or for that matter, automobiles, early examples of which were detested by some because they scared the horses.)

A similar progression may be occurring with regard to cloning. In the United Kingdom the Human Reproductive Cloning Act 2001 explicitly prohibited reproductive cloning. But then the Human Fertilisation and Embryology Act of 2008 repealed the 2001 Cloning Act, and allowed cloning and experiments on hybrid human-animal embryos. In the United States there are no federal laws that ban cloning, but some states do have such laws. The main argument against cloning is that the difficulties observed in cloning other animals makes it likely that there would be many failures in the creation of a living human clone and hence many disabled children. One can imagine at least two possible outcomes of the current ambivalence toward cloning. One is to increase the effectiveness of agricultural cloning until its success rate is very high. This would likely be followed by greater societal acceptance of the practice. The second possibility is to establish a line of tissues for individuals derived from their induced pluripotent stem cells. This would blur the line between stem cell therapy (routine for bone marrow transplants) and therapeutic cloning.

One of the most common objections people make about the prospect of human immortality is the unintended consequence of overpopulation. They also worry about long-lived individuals taking away jobs from younger people.

People have been worried about overpopulation ever since Malthus and his "Essay on the Principle of Population," which was published in 1798 (and in which he said, amid an outpouring of dire forecasts: "I happen to have a very bad fit of the tooth-ache at the time I am writing this"). Thomas Malthus died in 1834 worried about the world population of 1 billion people doubling every three hundred years. Currently at 7 billion and doubling every forty-seven years, we are now well beyond the situation that worried him. But instead of the global starvation and misery he envisioned, we have seen rises in wealth, standard of living, health, personal hygiene, and life expectancy. There is a reason for this: as economist Julian Simon once explained, "Resources come out of people's minds more than out of the ground or air. Minds matter economically as much as or more than hands or mouths. Human beings create more than they use, on average. It had to be so, or we would be an extinct species."

The world population has more than doubled since the beginning of the Green Revolution in 1943. Among the new technologies that have made this possible are pesticides, new irrigation strategies, synthetic nitrogen fertilizer, and mutant high-yielding varieties (HYV) of maize, wheat, and rice plants that are of shorter stature and use the new fertilizers. Since 1996 soybeans and other types of crops every year have been engineered to be resistant to herbicides, enabling a potent strategy potentially reducing erosion and the energy costs of plowing. However, now that weeds may be developing resistance to pesticides we need to be much more strategic in our planning.

The wide use of perennials may go even further in this direction. The return of our great American prairie made of deeply rooted perennial grasses would out-compete weeds, and greatly reduce erosion, irrigation, herbicides, and fertilizer. These perennials are similar enough to their annual growing season cousins, that developing a win-win perennial grain through genome engineering seems entirely feasible. Examples already under development include perennial rice and intermediate wheatgrass. But the land now must provide not only food but also the biofuels and other petrochemicals that were once cheaply available from underground. One solution, as we have seen, is engineered cyanobacteria, which can grow on marginal lands and use brackish or ocean water unsuitable for most conventional crops, possibly reclaiming land recently lost to desertification. And they can do so at photosynthetic efficiencies far higher than corn, switchgrass, and other favorite crops. Creative food technology might make cyanobacteria-derived cuisine taste the same as chicken or beef (or even better!), with consequent fifteen-fold reductions of energy use relative to animals together with a much lower risk of starting drug-resistant pandemics. The rapid diffusion of these technologies could lead to increased local production of simple and complex goods and hence reduced transportation costs.

But remember that there has been a worldwide shift from rural to urban life (worldwide, 3 percent urban in 1800 to 80 percent urban projected for 2050). In the course of this shift, the average number of children per family drops from an average near 7 to well below the break-even point of 2.1, and often as low as 1.2. This means that overpopulation may not be the problem that many people reflexively imagine it to be. Indeed, population may implode rather than explode, reversing the conventional wisdom on this topic.

The vision of a nearly immortal populace squelching the job prospects of youth is strongly reminiscent of Luddite concerns about machines taking over jobs from humans. Population implosion, coupled with increasing numbers of older, healthier citizens, and more women in positions of power, could have huge consequences for child rearing, consumer advocacy, philanthropy, and diplomacy. The assumption of ever expanding numbers of descendants and consequent fighting for their lands is hard to shake. With children a rarer resource, education may go from being among the lowest paying jobs to the highest. We may embrace much greater human diversity, not merely ancestry but vast spectra of personality, age, and intellectual capacities (e.g., an intentional increase in high-functioning autistics, bipolars, and ADHDs). This may require very specialized and highly trained parenting—well beyond the current random assignment of child to parent.

+1 YR, THE END OF THE BEGINNING, TRANSHUMANISM, AND THE PANSPERMIA ERA

Societal Risks and Countermeasures

৪৯

A popular theme in techno-thriller novels is the story of mutant microbes running amok and almost, but not quite, wiping out the human race. The *locus classicus* for this scenario is Michael Crichton's *The Andromeda Strain* (1969). Here a deadly microbe of extraterrestrial origin lands near the fictional town of Piedmont, Arizona, and the invading organisms slaughter the town's entire population, with the exception of two residents who mysteriously survive. The microbe responsible, the Andromeda strain of the title, mutates with each cell division and acquires new and even more destructive biological properties. The fate of humanity hangs in the balance, until a government scientist working inside a secret and secure biological containment facility somewhere in Nevada heroically carries out a last-minute save. The Andromeda microbes then depart for the upper atmosphere, whose lower oxygen content better suits their growth. (This story line seems to be a Michael Crichton specialty. His 2002 novel,

Prey, depicts a race of artificially alive nanorobots that escape from the lab, collect themselves into swarms, and then hunt down and kill people.)

Paranoia about exotic microbes is not confined to fiction, however. In real life it underlies the hostility toward, and temporary suppression of, the first generation of recombinant DNA experiments during the mid-1970s. It's what motivates panic reactions to genetically modified foods. And it's also what incites some knee-jerk criticisms of synthetic biology. "Scientists are making strands of DNA that never existed," claims Jim Thomas, a synthetic biology critic with the ETC Group, a technology monitoring organization based in Ottawa. "So there is nothing to compare them to."

Well, not exactly. Every two parents who conceive a child are creating "strands of DNA that never existed." Bacteria are constantly exchanging genes in a process called conjugation, thereby giving rise to yet more "strands of DNA that never existed." In fact, the creation of DNA that never before existed is a constant and pervasive feature of life on earth.

Despite the fear evoked by the idea of genetically modified organisms, those of the natural variety are hard to beat when it comes to posing serious threats to humanity. *Yersinia pestis,* the causative agent of bubonic plague, is estimated to have killed as many as a third to a half of all Europeans in the Black Death epidemic of the mid-fourteenth century. The bacterium made a comeback appearance in the Great Plague, another wave of annihilation that swept through the Continent in 1665–1666. The smallpox viruses, *Variola* major and minor, are thought to have caused more deaths than any other disease in human history, wiping out as many as 300 to 500 million people in the twentieth century alone. (The disease was eradicated in 1979.) Tuberculosis, malaria, cholera, AIDS—all are products of microbial agents of mass destruction that are natural in origin.

So it's not as if we have to look to genetically altered microbes to find models of successful microbial killing machines. Still, the question has to be faced whether the process of genetic engineering is likely to make existing microbes even more deadly, giving rise to one or more Andromeda strains that originated in biotechnology labs here on earth.

And if genetically modified microbes pose a threat, what about genetically enhanced human beings, so-called transhumans? It's a cliché that the human being is the most dangerous animal. Would a race of transhumans be even worse?

     ಜ ಜ ಜ

The term "transhuman" harnesses the prefix "trans," as in across or beyond, or acts as an abbreviation for transitional human. The transhuman occupies an intermediate stage between a normal biological human and one of the posthuman variety, a being whose capacities so far outstrip those of ordinary, everyday mortals as to constitute a new and separate species. The nomenclature is a bit imprecise, however, owing to the fact that the concept of improving human qualities to the point where they make up a species difference is a blend of ideas from science fiction, science, folklore and legend, and sheer crackpottery.

Nonetheless, transhumanism (sometimes symbolized as H+) has been taken seriously enough by some as to warrant attention and criticism. In 2004, for example, the magazine *Foreign Policy* published a special report, "The World's Most Dangerous Ideas." In it, Francis Fukuyama, a professor of international political economy at the Johns Hopkins School of Advanced International Studies, proposed the idea of transhumanism as one of the most dangerous and revolting products of those he referred to as "genetic bulldozers" (such "bulldozers" apparently being people who push their transhumanist agenda with lots of force).

This was puzzling inasmuch as the author acknowledged that transhumanism (which he defined as the attempt to "use biotechnology to make ourselves stronger, smarter, less prone to violence, and longer-lived") had a certain inherent logic and appeal. Part of the contemporary biomedical research agenda, he admitted, was focused on procedures and technologies that were designed as much to enhance ourselves psychologically and biologically as to cure illness; medical science presented us with a range of mood-altering drugs, prenatal genetic screening tests, and gene therapies, among other things. Moreover, the author also conceded that "the human race, after all, is a pretty sorry mess, with our stubborn diseases, physical limitations, and short lives."

Fukuyama, nevertheless, was an implacable foe of transhumanism, his reason being that "the first victim of transhumanism might be equality." By this he meant political equality, or equality before the law, not economic equality or uniformity of outcomes. He imagined that transhumans, with their heightened powers, better health, better looks, smarter minds, and longer life spans, would claim comparably outsize rights for themselves.

But by any standard this was an odd argument. There is already a wide variation in talents among members of the human race, but those with great intelligence, strength, and good health do not claim special rights for themselves on that account. Conversely, those with severe physical and/or mental disabilities are nevertheless accorded full human rights by the legal structures of enlightened democratic governments.

This is not to say that transhumanism presents no dangers whatever or that it entails no unintended or unwelcome consequences. A small number of individuals already exist who possess qualities that could be regarded as "transhuman" in a limited and qualified sense of the term. Rare double mutants in the myostatin gene (a.k.a. MSTN) have lean muscle and low body fat. Rare mutants in the LRP5 gene have extra-strong bones. Mutants in PCSK9 have 88 percent lower coronary disease. Double mutants in CCR5 are HIV resistant (see Chapter 9). Double mutants in FUT2 are resistant to stomach flu (e.g., noroviruses).

On the other hand, consider those blessed, or cursed, by the condition known as hyperthymesia, the ability to recall autobiographical events in extraordinary detail. First reported in the scientific literature in the journal *Neurocase* as "A Case of Unusual Autobiographical Remembering" (2006), the subject was "a woman whose remembering dominates her life. Her memory is 'nonstop, uncontrollable, and automatic.' [She] spends an excessive amount of time recalling her personal past with considerable accuracy and reliability. If given a date, she can tell you what she was doing and what day of the week it fell on."

The woman, who later revealed her identity as Jill Price, of Los Angeles, wrote a book about herself called *The Woman Who Can't Forget* (2008) in which she described her condition as if it were a plague: "My memories are like scenes from home movies of every day of my life, constantly playing in my head, flashing forward and backward through the years relentlessly, taking me to any given moment, entirely of their own volition."

The team of neurophysiologists who examined her reported that Price, "while of average intelligence, has significant deficits in executive functions involving abstraction, self-generated organization and mental control," as well as "obsessive-compulsive tendencies" and more. Her superior memory, in other words, was counterbalanced by deficits in other areas.

Others with prodigious memories also have exhibited anomalous capacities elsewhere. Solomon Shereshevsky, a Russian journalist with a nearly photographic memory, also displayed synesthesia, a condition in which the stimulation of one of the five senses produced a reaction in one or more of the rest. Hearing a musical tone, for example, caused him to see a certain color, whereas the sensation of touching an object gave rise to a taste sensation.

Conventional wisdom, going back to the ancients, has it that genius and madness are often coupled in the same individual. In its less extreme form, this is the idea that those who are exceptionally talented in one area suffer compensating defects somewhere else in their personalities. And so we have the standard examples of Vincent van Gogh hacking off part of his ear and later committing suicide; Ludwig van Beethoven with his slovenly personal habits, absentmindedness, delusions of royalty, plus persecution fantasies and assorted other manias; and Isaac Newton, who was obsessively secretive and paranoiac, and who had fixations with astrology, alchemy, and the color red.

Among more run-of-the-mill geniuses, personal idiosyncrasies amount to mere quirks. Paul Dirac, for instance, the notoriously tight-lipped and taciturn physicist, once explained his uncommunicativeness: "My father made the rule that I should only talk to him in French. He thought it would be good for me to learn French in that way. Since I found that I couldn't express myself in French, it was better for me to stay silent than to talk in English. So I became very silent at that time."

Makes sense to me. Still, the question remains whether creating a race of transhumans would unwittingly produce a population burdened with a variety of unknown but severe dysfunctions or handicaps. And if we run

that risk with benign and well-intentioned people (or even with benevo-
lent amplified people), how much worse would be the potential for damage
wrought by an actively malicious supergenius? After all, the power of individuals is growing. In ancient times one person might murder a couple
of enemies with a rock. Later, he could destroy a village with fire. Today a
small team might kill millions of people with nuclear, chemical, or biological warfare agents. In the future, could a single transitional human
having a bad day (a Columbine teenager, for example) release a doomsday
WMD?

♬   ♬   ♬

Some of these same questions also arise with respect to synthetic genomics itself, and their relevance is especially clear in the case of iGEM,
whose avowed purpose is to make biological engineering "easy." Consid-
ering the numbers of college undergrads, and even high school students,
engineering novel organisms with special properties and engaging in
competitions for best design, best engineered biological part, and so on,
there are plenty of opportunities for things to go wrong in a big way,
whether by accident, deliberate misuse, or through the normal mutation
and evolution of organisms.

David Donnell, adviser to the Citadel's iGEM team, had doubts about
the safety of engineering *E. coli* for appetite control. The idea was that the
bacterial population would oscillate around some fixed mean that was sufficient to allow steady production of the appetite-suppressing peptide PYY.
Supposedly, the supply of microbes would be kept in check by an engineered quorum-sensing circuit that would regulate gene expression in response to fluctuations in cell population density levels. Researchers at
Duke had shown how quorum sensing could be used to limit the size of a
bacterial population, and Team Citadel hoped to wire this same ability
into their microbes. But could such quorum-sensing circuitry be made
fail-safe?

"I'd like to say that we should be able to build in sufficient fail-safes that
would allow us to pull the plug on the engineered bacteria if they got out

of hand," David Donnell said. "But of course, with a thick layer of oil sludge coating the ocean bottom in the Gulf of Mexico following the failure of a whole series of fail-safes designed to prevent oil spills by drilling rigs, I am somewhat skeptical of our ability to design a fail-safe fail-safe."

There are no fail-safe fail-safes in biological lab work. Laboratories that handle natural microorganisms (including pathogens) that are expressly designed to prevent mishaps from occurring and are staffed by highly trained and experienced professionals have nevertheless had numerous accidents. In *Biological Safety: Principles and Practices*, two biosafety experts, Diane O. Fleming and Debra Long Hunt, reviewed accounts of biolab accidents published in scientific journals from 1970 to 2004. Over this thirty-four-year period there were a reported 1,448 symptom-causing infections resulting in thirty-six deaths. And as for the possibility of a high-level expert working in a well-equipped biosafety lab suddenly going berserk and wreaking havoc, recall Bruce Ivins, the perpetrator of the 2001 anthrax letter attacks. Ivins worked at USAMRIID, the US Army Medical Research Institute for Infectious Diseases, with the mission of protecting people from diseases like anthrax, not spreading them.

Facts like these have prompted synthetic biology researchers to think critically about safety and security. (The concepts of safety and security are not the same. *Safety* involves preventing accidents with hazardous organisms, especially the unintentional release of an engineered microbe into the environment, where it could have unknown but potentially damaging effects. *Security*, by contrast, refers to preventing the deliberate misuse of engineered organisms, whether by rogue states, terrorist groups, or lone agents working by themselves.) For all the benefits it promises, synthetic biology is potentially more dangerous than chemical or nuclear weaponry, since organisms can self-replicate, spread rapidly throughout the world, and mutate and evolve on their own. But as challenging as it might be to make synthetic biology research safe and secure within an institutional framework such as a university, industrial, or government lab, matters take a turn for the worse with the prospect of "biohackers," lone agents or groups of untrained amateurs, working clandestinely, or even openly, with biological systems that have been intentionally made easy to engineer. The problem with making biological engineering techniques easy to use is that it also makes them easy to abuse.

In 2008 Markus Schmidt, an adviser to iGEM and a member of the Biosafety Working Group at the Organization for International Dialogue and Conflict Management in Austria, published a paper, "Diffusion of Synthetic Biology: A Challenge to Biosafety." In it he portrayed iGEM's attempt to make bioengineering easy as a "de-skilling" of biotechnological processes and as a "domestication of biology [that] could easily lead to unprecedented safety challenges." In the extreme case, he said, "imagining a world where practically anybody with an average IQ would have the ability to create novel organisms in their home garage without adhering to a professional code of conduct, filing a reporting system, and lacking a sufficient biosafety training is a thrilling thought."

In the early years of synthetic biology, researchers spoke of "do-it-yourself biology" as essentially a metaphor referring to the prospect of amateurs creating organisms in their kitchen at some unspecified time in the future. "Garage biology," likewise, was a jocular term of abuse. A decade or so later, however, those possibilities had become realities, sooner than many of us would have thought. The website biohack.sourceforge.net, for example, offers an "open, free synthetic biology kit [that] contains all sorts of information from across the web on how to do it: how to extract and amplify DNA, cloning techniques, making DNA by what's known as oligonucleotides, and all sorts of other tutorials and documents on techniques in genetic engineering, tissue engineering, synbio (synthetic biology), stem cell research, SCNT, evolutionary engineering, bioinformatics, etc."

And in 2008, Jason Bobe, director of community outreach for the Personal Genome Project, and Mackenzie Cowell, a web developer, formed an online do-it-yourself biology discussion group calling itself DIYbio: An Institution for the Amateur Biologist (diybio.org). On May 1, 2008, about twenty-five hardcore DIYbio members got together in a back room of Asgard's Irish Pub in Cambridge, just a few blocks from MIT, and discussed topics such as: Can molecular biology or biotechnology be a hobby? Will advances in synthetic biology be the tipping point that enables DIYers and garage biologists to make meaningful contributions to the biological sciences, outside of traditional institutions? Can DIYbio.org be the Homebrew Computer Club of biology? Good questions!

Two years later there were local DIYbio communities all over the globe, including several in the United States and Europe, three in India, and three more in South America. In addition, there were about 2,000 subscribers to the DIYbio mailing list. With this kind of rapid ideational diffusion, it was not long before agents from the FBI's Weapons of Mass Destruction Directorate were showing up at DIYbio meetings, which were openly announced on the group's website.

🙛   🙛   🙛

In 2009 the inevitable finally happened: garage biology came of age. Technically, the first garage biology operation began in 2005, when physicist Rob Carlson set up a private one-man lab in the garage of his Seattle home. Carlson had learned molecular biology techniques while working at Sydney Brenner's Molecular Sciences Institute in Berkeley, California. Eventually he realized that if he could put together parts of different protein molecules in a certain way, then he might have a commercial product on his hands. And so with reconditioned micropipettes and a used centrifuge bought on eBay, plus the usual array of lab glassware and other machinery and instrumentation, he started working nights and weekends as a sort of biotech hobbyist. His garage lab, Carlson joked, was "half start-up company and half art project."

Four years later, John Schloendorn, who had a doctoral degree in molecular biology from Arizona State, and Eri Gentry, a Yale economics grad who was his business partner, founded a biotech lab in Gentry's garage in Mountain View, California, and started doing paid anticancer research. The idea for the garage lab arose after a friend of theirs died of esophageal cancer. Schloendorn knew about several lines of evidence suggesting that a natural immune response to cancer existed in some people and he was brash enough to think that, working by himself in his own private lab, he could play a direct role in advancing the scientific understanding of such immunity. With $30,000 (much of which was Schloendorn's own money, the rest coming from Gentry and donations from friends), Gentry started outfitting the place. By shopping for used equipment at liquidations and auctions, as well as on eBay and Craigslist, she managed to purchase an estimated million dollars worth of machinery, including a custom-made clean bench, inverted phase contrast microscope, incubators, centrifuges, a fluorescence microplate reader, and so on.

They named the lab Livly. By the summer of 2010 Livly had received a round of seed funding from ImmunePath, a Silicon Valley start-up aimed at curing cancer with stem cell immunotherapy. Gentry, meanwhile, inspired by the DIYbio movement, investigated the idea of creating a community lab in the San Francisco Bay area. The idea was to open the lab to interested citizen-scientists who, for a monthly membership fee of about $200, could perform their own molecular biology experiments, everything from DNA sequencing to the rite-of-passage project of inserting green fluorescent protein genes into bacteria. If successful, it would be a case of wresting science from its conventional setting in industry and academia and returning it to something like the days of the gentleman-scientist-scholar who was not formally connected with any established organization: people such as Newton, Darwin, Alfred Russel Wallace, Benjamin Franklin, and others.

Gentry and two partners came up with the name BioCurious (biocurious.org) and started a fund drive on the website Kickstarter.com ("A New Way to Fund and Follow Creativity"), where inventors, entrepreneurs, and dreamers of every stripe could post their wild schemes and pet projects and ask for money to fund them. BioCurious announced an initial goal of $30,000. The partners were soon oversubscribed, almost overwhelmed, with 239 backers pledging $35,319. In the fall of 2010 Gentry and her partners were looking to lease 3,000 square feet of industrial space in Mountain View, but in the end settled for a 2,400 square feet in Sunnyvale, calling it "Your Bay Area hackerspace for biotech." In December 2010, meanwhile, another DIY biohacker lab, Genspace, opened in Brooklyn, New York. The founders referred to it as "the world's first permanent, biosafety level 1 community laboratory" (genspace.org). Many others soon followed, in the United States, Canada, Europe, and Asia.

With free synthetic biology kits, DIYbio, Livly lab, BioCurious, Genspace, and others, the synthetic biology genie was well and truly out of the bottle.

ℱ҄   ℱ҄   ℱ҄

That did not mean we were headed for some sort of synthetic biology holocaust, Armageddon, or meltdown, however. For one thing, despite efforts by iGEMites to make biological engineering "easy," it is still reasonably difficult to design and implement biological circuitry that actually works (much less works reliably and exactly as planned). Biological systems are complex, "noisy," and susceptible to mutations, evolution, and emergent behaviors. For these reasons their operations are full of surprises. A random change to any given genome is more likely to weaken the organism than to strengthen it, and the same is often true of changes that have been carefully designed and engineered in advance.

For another, in parallel with the development of the biohacker, DIYbio, and garage biology movements (and in fact slightly predating them), many of those doing hands-on synthetic biology work had written white papers, position papers, opinion pieces, and had participated in conferences, study groups, and other organized attempts aimed at dealing with the risks and dangers posed by engineering life. In 2004, for example, I wrote "A Synthetic Biohazard Nonproliferation Proposal." Here I advanced two main ideas for enhancing the safety and security of synthetic biology research. The first was that the sale and maintenance of oligonucleotide (DNA) synthesis machines and supplies would be restricted to government-licensed nonprofit, government, and for-profit institutions; in addition, the use of reagents and oligos would be tracked, much after the manner of nuclear materials. Second, orders placed with commercial oligo suppliers would be screened for similarity to known infectious agents. While unpopular back then, today many suppliers now do this.

Admittedly, both measures had limitations. Restricting the sale of DNA synthesis machines to licensed or "legitimate" users would be easier to apply to future sales and do little to restrict the use of the machines that were already out there. Also, it's an open question how stringently enforced such licensing could be. Further, the proposal to screen orders placed with DNA synthesis companies would not affect firms that, for proprietary, competitive, or other reasons, did their DNA synthesis with their own equipment, in-house, unless those instruments fell under the same regulations.

Still, further measures to keep people safe and to keep engineered organisms under control have been proposed by myself and others. One is that synthetic biology research be done in physical isolation, employing the same safeguards that are routinely employed in biosafety labs: safety cabinets, protective clothing and gloves, plus face protection. (But as we have seen, accidents happen even in biosafety labs.) Another is to build one or more self-destruct mechanisms into engineered organisms so that they would die if they escaped from a lab. They could be made dependent on nutrients found only in a laboratory setting, or they could be programmed with suicide genes that would kill the organism after a certain predetermined number of replications, or in response to high-density levels of the organism, or even in response to an externally generated chemical signal. Farther into the future, you could also base the organism on a genetic code different from the one used by natural organisms. Such a code change, I have argued, or the introduction of "mirror life" (see Chapter 1) would mean that such organisms would not be recognizable to natural organisms, and therefore would be unable to exchange genetic material with them.

Further safety measures have been proposed by others. In their 2006 piece, "The Promise and Perils of Synthetic Biology," Jonathan Tucker and Raymond Zilinskas suggested that before a genetically engineered organism was released to the wild, it first be tested in a microcosm, a model ecosystem ranging in size from a few millimeters to fifteen cubic meters (about the size of a one-car garage), and then to a "mesocosm," a model system larger than fifteen cubic meters. "Ideally," they wrote, "a model ecosystem should be sufficiently realistic and reproducible to serve as a bridge between the laboratory and a field test in the open environment."

Synthetic biology needs a suite of safety measures comparable to those that we now have for cars. Modern cars represent powerful technology, and so we require licensing, surveillance, safety design, and safety testing of both the cars and their drivers. Despite the complex technology of today's automobiles and traffic management, the average citizen is expected to be able to pass a written and practical licensing exam. In addition, we enact numerous surveillance procedures, including radar monitoring of speed, license plate photos, timing speed between tolls, annual inspections, visual checks for erratic behavior, weighing trucks, and checking registration papers. The cars are designed for safety, including anti-hydroplaning tires; antilock brake systems; seat belts; shoulder harnesses; front, side, and supplemental air bags; and so on. In addition, cars are tested using actual cars and synthetic humans—originally cadavers, and later increasingly realistic crash dummies. But even with all these systems, checks, devices, and procedures in place, there are still about 40,000 automotive deaths annually in the United States alone.

We could similarly require licensing for all aspects and users, even DIY-Bio; computer-aided surveillance of all commerce; designing cells that self-destruct outside of the lab; and rigorous testing of what would happen if the cell escaped from the lab by bringing the outside ecosystem into the lab in a secure physical containment setting.

Regulations, however, can be circumvented by anyone who is sufficiently determined to evade them. In other words, security is far more difficult to achieve than safety. This point was made repeatedly by the authors of the 2007 report, *Synthetic Genomics: Options for Governance.* The document was the fruit of an exhaustive two-year study funded by the Alfred P. Sloan Foundation. The study involved eighteen core members (including Drew Endy, Tom Knight, Craig Venter, and myself) and three institutes: the J. Craig Venter Institute, MIT's Department of Biological Engineering, and the Center for Strategic and International Studies.

Our final report advanced many policy options along the lines of those already mentioned. We made no bones about the fact that their "security benefits would be modest because no such regime could have high confidence in preventing illegitimate synthesis." DNA synthesizers, after all, were relatively small (desktop-size) machines, easy to acquire and hide from view. Even if the registration of synthesizers were legally required,

the policy would be difficult to enforce because it would be virtually impossible to ensure that all existing machines had been identified and incorporated into the registry. Furthermore, DNA synthesis machines can
be built from scratch, can be stolen, and can be misused at "legitimate" institutions by someone posing as benign and genuine while nevertheless
engaging in illicit activity (the Bruce Ivins paradigm).

The group tackling options for governance considered the difficulty of
synthesizing several pathogenic viruses, including the 1918 influenza
virus, poliovirus, Marburg and Ebola viruses, and foot and mouth disease
virus. (Foot and mouth disease affects only certain hoofed animals such
as sheep and cattle, but it is highly contagious and could trigger the wholesale loss of herds that in turn would entail carcass removal and deconta-
mination costs. An outbreak would destroy consumer confidence, cripple
the economy, and provoke trade embargoes.) Of these viruses, we classi-
fied two of them—poliovirus and foot and mouth disease virus—as easy
to synthesize by "someone with knowledge of and experience in virology
and molecular biology and an equipped lab but not necessarily with advanced experience ('difficulty' includes obtaining the nucleic acid and
making the nucleic acid infectious)."

In the end, we found no magic bullets for absolutely preventing worst-
case scenarios, no fail-safe fail-safes, but in my opinion the measures we
proposed are worth implementing anyway since their costs are low, the
risks high, and their effectiveness would be measureable, and subject to
improvement.

ళ   ళ   ళ

To be on the safe side, then, why not prohibit the entire enterprise, or at
least the riskiest parts of it? Given the amount of information, machinery,
and engineered organisms that already exist in the world, total prohibition
would be unrealistic even if it were desirable, which it is not: synthetic genomics offers too many benefits in comparison to its risks. And there are
powerful arguments against prohibiting even a subset of experiments or
research directions that might be considered relatively dangerous. The first
is that prohibitions mostly don't work; instead, they merely drive the pro-

hibited activity underground and create black markets, or clandestine labs and lab work that are more difficult to monitor and control than open markets and open laboratories. The second is that they also produce a raft of adverse unintended consequences, many of them foreseeable.

The classic case, of course, is Prohibition, which was enacted in 1920 by the Eighteenth Amendment outlawing "the manufacture, sale, or transportation of intoxicating liquors" within the United States. The law stopped none of that activity and instead created a huge network of illegal alcohol production, distribution, and transportation, including massive smuggling across the Canadian border. At one point there were 20,000 speakeasies in Detroit alone (one for every thirty adults). Millions of formerly law-abiding citizens suddenly became habitual lawbreakers. Many drinkers were poisoned by poorly made bootlegged liquor. The net effect, in sum, was to increase crime, violence, and death.

The other major prohibition story of our time is the war on drugs, which has created a cottage industry of illegal drug production, transportation, distribution, and sale in the United States and abroad. It also fosters ingenious methods of drug smuggling, including the use of false-paneled pickup trucks, vans, and tractor trailers, and the building of air-conditioned tunnels under the US-Mexican border. But the ultimate high-tech drug-running innovation was the home-built submarine (called narco subs in the trade), used to move large amounts of cocaine underwater. In July 2010 Ecuadorean police discovered a so-called supersub in the jungle, a seventy-four-foot (23-meter), camouflage-painted submarine that was almost twice as long as a city bus, was equipped with diesel engines, battery-powered electric motors, twin propellers, and topped by a five-foot conning tower. With a crew of four, the sub had a range of 6,800 nautical miles, and in its cargo hold could carry nine tons of cocaine, worth a total of about $250 million. Jay Bergman, the top US Drug Enforcement Agency official in South America, said of the sub, which he praised as "a quantum leap in technology," that "it poses some formidable challenges."

Still, it is possible to outlaw entire technologies. In 2006 Kevin Kelly, the former editor of *Wired* magazine, did a study of the effectiveness of technology prohibitions across the last thousand years, beginning in the year 1000. During this period governments had banned numerous technologies

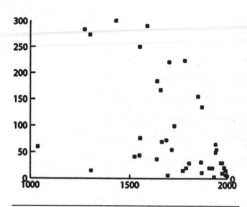

Figure Epilogue Kevin Kelly's chart of the duration of a technology prohibition plotted against the year in which it was imposed.

and inventions, including crossbows, guns, mines, nuclear bombs, electricity, automobiles, large sailing ships, bathtubs, blood transfusions, vaccines, television, computers, and the Internet. Kelly found that few technology prohibitions had any staying power and that in general, the more recent the prohibition, the shorter its duration.

"Prohibitions are in effect postponements," Kelly concluded. "You might be able to delay the arrival of a specific version of technology, but not stop it. If we take a global view of technology, prohibition seems very ephemeral. While it may be banned in one place, it will thrive in another. In a global marketplace, nothing is eliminated. Where a technology is banned locally, it later revives to global levels."

Even if a technology is banned globally, as for example under the terms of a legally binding international treaty, the ban will not necessarily stop its development. In 1972 seventy-nine nations signed the Convention on the Prohibition of the Development, Production, and Stockpiling of Bacteriological and Toxin Weapons, more popularly known as the biological weapons convention (BWC). But in 1996, more than twenty years after the weapons convention came into force, US intelligence sources claimed that twice as many countries possessed or were actively developing biological weapons as when the treaty was signed. Most of the violators of the convention, including the Soviet Union, India, Bulgaria, China, Iran, Cuba, Vietnam, and Laos, were also signatories to the convention.

The moral of the story is that prohibitions are generally ineffective and counterproductive, and have negative unintended consequences. There is no reason to think that a prohibition would halt the development of synthetic genomics, although it might slow down the pace of progress—and the potential benefits that unimpeded progress would have brought.

ℬℴ ℬℴ ℬℴ

In general, concerns about a new technology arise mainly during the transition to it. The year 2010 marked the 200th anniversary of the Luddite response to the industrial revolution. It was also the thirty-fifth anniversary of the recombinant DNA moratorium. Ten years earlier a few deaths in the first gene therapy trials drastically reduced funding for this promising new field. But the industrial revolution that the Luddites tried to prevent in 1811 has brought us enormous benefits.

Travel speeds could be considered one of our first transitional human traits—going from natural long-distance rates of 10 km/hr to 1,000 km/hr in passenger jets, to 26,720 km/hr in orbit. Other technologies have radically increased our ability to sense and interact with the universe around us.

	Past	Current
Input		
Visible light	4 to 7×10^{-5} meters	10^{-12} to 10^6 meters
Hearing	10 to 20,000 Hz	10^{-9} to 10^{12} Hz
Chemosenses	5 tastes, 1,000 smells	Millions of compounds
Touch	3,000 nm	0.1 nm
Heat sensing	200 to 400 K	3 to 10^5 K
Midput		
Memory span	20 years	5,000 years
Memory content	10^9 bits	10^{17} bits
Cell therapy	0	Many tissues
Heat tolerance	270 to 370 K	3 to 10^3 K
Output		
Locomotion	50 km/h	26,720 km/h
Ocean depth	75 meters	10,912 meters
Altitude	8×10^3 meters	3×10^9 meters
Voice	300 to 3,500 Hz	10 to 20,000 Hz

Our technologies have already given us more than a few transhuman qualities, and the trend toward transhumanism is only likely to accelerate in the future. Kurzweil's Law of Accelerating Returns notes that the progress of certain technologies is not linear but exponential (see Figure 7.1), and Kurzweil himself holds that future technological change will be so rapid and profound that it will constitute "a rupture in the fabric of human history."

Of course a healthy dose of scientific skepticism is always prudent in the face of such extrapolations. For example, an event or series of events may occur that suddenly derails the rate of accelerating returns. The anti-predictionist Nassim Nicholas Taleb, in his book *The Black Swan*, presents a vivid example of how a prolonged and steady trend may come to an abrupt halt unexpectedly: "Consider a turkey that is fed every single day. Every single feeding will firm up the bird's belief that it is the general rule of life to be fed every day by members of the human race ... On the afternoon of the Wednesday before Thanksgiving, something *unexpected* will happen to the turkey. It will incur a revision of belief."

On a more scientific level, physicist Stephen Wolfram has claimed that certain systems are so complex that it's inherently impossible to predict their future behavior reliably by any means. He regards such systems as "computationally irreducible." It can be argued that the body of technologies we have today forms a complex system that is computationally irreducible, especially given the fact that these technologies are products of millions of human minds, free agents, and innumerable decisions. If that complex, dynamic network is in fact computationally irreducible, then in Wolfram's words, "there can be no way to predict how the system will behave except by going through almost as many steps of computation as the evolution of the system itself."

But even if the rate of technological progress, including genomic technology and the march toward transhumanism, is not knowable in advance, it is at least within human control. And that should be a comforting thought.

ﬁ   ﬁ   ﬁ

At the end of such a great story you may ask, What's next? The safest way to make a bet about a future event is to heavily influence it. Like many great stories, the story of the genome includes a moral, a prescription for the future. Twenty years, the length of time we've been able to read the language of DNA, seems ridiculously short compared to the long saga of the genome itself. But twenty years seems like an eternity now that key technologies (like electronics) are changing exponentially at a 1.5-fold per year rate, or even at the 10-fold per year speed at which improvements are being made in DNA reading and writing. Whether passive or active, we can study the future by projecting the consequences of taking either of two branches at the many forks in the road ahead—and taking each to its logical extreme.

1. **Natural versus exotic:** How will we change ourselves individually? This can be done without cloning. Using ever more complex consumer devices and by engineering our adult biology, we could all become more alike (e.g., the best of our ancestors or contemporaries) . . . or more radically different from each other. With so many decision forks confronting us, and with so many options ahead of us conferred by advanced technologies, we cannot be content simply to ask, What *can* we do? No. When we're faced with such myriad possibilities, we must go one step further and ask, What *should* we do? While some may object that this is not a scientific question, yet an attempt to answer it is a neural and an evolutionary process (hence subject to scientific inquiry), and anyway, it's a question forced on us by the breakneck progress of science itself.

So, how should we prioritize our research and life goals? Without goals and measures of progress, we are amoral, apathetic, and wasteful. We could join George Mallory, who justified a huge and risky effort with the phrase "Because it's there"—referring to Mount Everest where his body was lost, frozen in ice, for seventy-five years. But this is to blunder along blindly. In a similar vein, we could say that "pure science" payoffs aren't predictable, meaning that it's futile even to try to plan ahead. But these are clearly cop-outs. We might say that our mission should be to maximize happiness or reduce suffering, but some might respond that suffering and death are natural and therefore desirable, or required for radical change and therefore good for us, at least as a species.

Perhaps there is no one thing that all of us should do collectively, but I would like to offer my own personal take on what at least some of us should be doing, albeit on a vast level. As a general goal I propose that, as a minimum, we ought to avoid the loss of all intelligent life in the universe. A variety of models and measures indicate that meteor impacts have caused massive planet-wide extinctions and are likely to do so again. What should we do? We should develop equipment for rapidly detecting and deflecting such events and/or moving some of our civilization out of the way, and off the planet. Clearly, technological stagnation; economic depression; exhaustion of key nonrenewable resources; conventional warfare; nuclear, biological, or chemical terrorism; environmental waste; pandemics; and various combinations of these could interfere with our ability and our will to deflect extraterrestrial threats—as well as constituting potential existential threats in their own right.

What should we do? Doing nothing, or doing what is traditional or natural, is not even close to a recipe for survival. If we chose tomorrow to behave in the way that our primitive ancestors did, nearly all of our 7 billion humans would die.

The genome should become not just the genome of one lonely being or one planet. It should become the genome of the Universe.

2. Staying at home versus jet-setting: The stay-at-home option could be driven by the desire to avoid energy waste, the terrorist threat, the tally of 1.2 million global transportation accidents per year, and the prospect of pathogen exposure. The same choice could be made even more attractive by improvements in virtual reality, 3-D displays with gigapixel teleconferencing, and easily engaging other senses like tactile (haptic) feedback. The ultimate jet-set option, by contrast, would be the rocket ride, escaping the inevitable collision of a stray rock with the earth, while adding new challenges of choosing a destination, ensuring radiation protection, and maintaining diversity, energy, and sanity along the way.

3. Personal versus generic: Do-it-yourself versus outsourcing. Examples of the former include the use of home solar chemistry and solar power, 3-D printers capable of producing complex objects on demand, DIY personal medicine, personal genomes, highly personalized solar foods, and personalized movies. At Harvard's Wyss Institute for Biologically Inspired Engineering, a regenerative medicine project is under way with the goal of reprogramming adult cells and then multiplexing the printing of scaffolding at very high resolution. This huge step toward individualized tissue engineering ultimately could seem no more intimidating than tattooing is today, and possibly less so if it's easily reversible. Many of us already get stem cell transplants as part of routine medical care, for example, both blood and skin. If we are up for a transplant anyway and are presented with the options of cells from an adequately matched donor (probably needing immune suppression) or our own cells engineered to be more resistant to our genetic, microbial, or cancer-based diseases, many of us would opt for the new, improved cells, especially since our ability to do quality assurance has improved so much recently.

The goal of optogenetics, the process of reading and writing to brain cells with light, is to enhance our ability to get measurements of neural activity from electrical to optical methods for the purpose of understanding and controlling how our brain operates on the cellular level. However, it's not obvious how much more scalable this might be for brain cells stimulated by multi-electrodes, which currently number in the dozens per brain. We need a way of getting a recording device for each of our 60 billion neurons (40,000 per cubic mm). This potential logjam argues for an alternative to hardwiring, the use of a wireless network or DNA recording of neuronal activity, rather than electrical or optical cables, perhaps employing 10 micron chips disguised as white blood cells that naturally penetrate the blood-brain barrier. This is the logical extrapolation of the demand for growing the bandwidth of the interface between our electronic computers and our biological (brain) computer and catching pathological brain events (ministrokes, onset of psychiatric disorders, etc.) as early as possible.

4. Priceless versus worthless: The cost of materials today ranges from $0.1 per kg for wood to $4 trillion per kg for certain pharmaceuticals (reimbursable by health insurance). With revolutions in smart materials and molecular engineering, all materials and objects could be reduced to the range of $0.2 per kg (electronics, clothes, foods, cosmetics, and so on)—or people could spend more and more for less and less via clever branding, copyright and patent laws, elaborate licensing and regulatory schemes, and the like. Or is there a way of artfully combining and integrating all of the above?

5. Rich versus poor: A growing global middle class can shrink the gap between rich and poor. As our species has moved from 2 percent of the population living in cities in ancient days to 80 percent in cities in the near future, the number of births per family has dropped from 8 to 1.2 (whereas 2.1 is needed for break-even). The fraction of the world living in extreme poverty and experiencing violence is decreasing; several technologies combined with earth shrinking and flattening (due to transportation, telecommunications, and trade) could accelerate this trend. Access to information and hence personalization means more resources available for further improvements.

6. Privacy versus publicity: Even if governments ever stay clear of spying on their own citizens or others, we will be increasingly motivated to share data about ourselves and everything that we see and hear—for reasons of security, personalization, and entertainment. Privacy is a relatively new social phenomenon. When our ancestors died in the tiny village of their birth, surrounded by relatives, there were no secrets. Privacy versus security is in fact a false dichotomy. A third option, lowering the need for secrets, seems to be growing in momentum. Consider that over the past decades the number of people concealing their psychiatric status, sexual orientation, STDs, cancer, and salary has shrunk for a variety of reasons. Some of these characteristics seem less susceptible to extortion or scandal than in the past. People share more now because new technologies make data more accessible (2012 Google versus 1950 private eye), or make the process of sharing exciting, fun, and chic (Facebook). Technologies also make sharing more personally valuable, for example, discussing which new drugs to take for cancer, AIDS, or depression. In forums such as PatientsLikeMe and PersonalGenomes.org, individuals communicate information to benefit people around the world.

So in addition to economic incentives to voluntarily give up privacy, attempts to sell secrecy result in false security. The memory hole of Orwell's *1984*, a purposefully disingenuous illusion of information security, is not so fictional (e.g., deleted personal web searches are restored in crime investigations). Add to that human error and willful individuals and teams, and we see strong arguments against dark secrets with nowhere to hide. Secrets are symptoms—not demanding a better bandage but a treatment for the underlying disease causes. AIDS created an activism that led to open discussion and reduced the stigma associated with sexual preference. Even "essential" secrets maintained by police and war fighters are symptoms of a failure of diplomacy and a failure of technology policy to provide a decent life for all.

There may be a trend toward less violence with global improvements in the standard of living or education—and as we learn more about the underlying biology of violence. What will we see as anachronistic as we look back from a few decades hence? Will it be the wimpiness of our security or the prevalence of our secrets?

Yes, evolution has evolved deceptions, for example, camouflage or mimicking poisonous species. Could those sometime serve the common good? Maybe. But evolution also gave us malaria, HIV, tuberculosis, and smallpox, among other things. Perhaps better than deception would be the notion that we need to separate ideas/memes/cultures for long enough to test them—and then recombine the parts we like best. But separation could be done without deception.

Let's say that we had strong artificial intelligence. Would we prefer to be able to read computers' minds or teach them how to lie to us and each other? Hopefully we will want the former. We would prefer to have them explain and justify their decisions rather than just strut off while taunting "It's for me to know and you to find out!" We generally engage the rule

"Never hide information from the programmers." The ability to check the robot's logic when a bug is found is valuable.

We can imagine scenarios where a person might not want to know something—but that should be their choice, not the choice of some paternalistic data-hoarding computer. Software should help people decide if information is useless, harmful, or helpful to them (before they see the details).* Deception and secrecy (as typically implemented) do not have such choices as top priorities. What kinds of information would hurt people's feelings for no purpose? If you tell me I'm ugly, that could hurt my feelings because society favors handsome people. If I need a shave in order to succeed, then it helps me more hearing about it now rather than a decade from now after I've been passed over in the marketplace of life.

You could hurt me by telling me that I invested stupidly, or had cancer, or was conceived out of wedlock, or someone was insulted by my narcolepsy or by false rumors. Overall, the openness gives me a better chance of nipping any such rumors in the bud.

We may decide to dispense with our current vast and destructive mechanisms for secret keeping (and deception). Cases in which new societal trends favoring secrecy override the need to know (or want to know) of one or more individuals may become rare, as they were centuries ago.

7. Will we become a new species? This new species is sometimes called *Homo evolutis*, posthuman, transhuman, parahuman, or H+. It seems likely that legal, moral, and ethical concerns will loom larger and sooner due more to selection than to speciation—and due more to mixing of species than to isolating one species from another. We are already using parts from hundreds of previously separate species/kingdoms in human-induced pluripotent stem cells, DNA enzymes from fungi that promote genetic recombination, transcription factors from *Xanthomonas*, green fluorescent protein (GFP) from jellyfish, enzyme genes from plants to make essential fatty acids, and so on down a long list (see Chapter 2). The interspecies

---

* Still, the revelation in December 2011 that some smartphones track people's keystrokes and web pages visited shocked and dismayed many users.

barrier is falling as fast as the Berlin Wall did in 1989. Not just occasional horizontal transfer but massive and intentional exchange—there is a global marketplace for genes. Not the isolating effect of islands or valleys result- ing in genetic drift and xenophobia, but a growing addiction to foreign gene products, for example, humans "mating" with wormwood for anti- malarial drug precursor artemisinin, and with *Clostridium* for Botox. Maybe instead of the genus name *Homo*, we should adopt the genus name of our chimpanzee cousins, *Pan* (derived from the mischievous Greek god of that name, but note also the prefix "pan" for "all-inclusive," in this con- text, giving us a double entendre pointing to our genus increasingly using bits of DNA from the whole biosphere).

One fundamental question is, How much does speciation need isola- tion, for example, the type of ecological isolation between finch popula- tions on Galapagos that, in part, led Darwin to develop the theory of evolution by natural selection? The earth used to be large relative to mi- grations. Humans died within a mile of their birth (one 25,000th of the earth's circumference). Now ordinary people commute 8.8 million miles in twenty-seven years. In addition to greatly enhanced travel, Stewart Brand would say that progressive urbanization means that by 2050, 80 percent of the population will live on just 3 percent of the land. So we will be literally running into each other on foot (without need for jets), as we are already doing in major cities. Such extreme population density will radically reduce isolation as a speciation driver. We would need a new way to achieve geographical isolation. Even space colonies may not do it if we can exchange ideas and DNA recipes at the speed of light.

Furthermore, the trend of relaxing rules against intermarriage seems to be taking us in the opposite direction from what's needed for speciation and for the evolution of a new human species. In *Loving v. Virginia* (1967), for example, the Supreme Court declared Virginia's anti-miscegenation law unconstitutional; the South African Immorality Act of 1927 (which banned sexual relations between whites and blacks) was repealed in 1985; and so on. Nevertheless, one way to achieve isolation would be for some of us to get far from our mother planet (a good survival idea in any case). Another way would be to design a species that is instantly incompatible with the current edition of *Homo sapiens*. Yes, I know: it's hard enough to design the properties of a truly new protein, much less a truly new genome! Still, one exceptional example is that of mirror life. Mirror people, as we have seen, would be immune to nearly all current pathogens, and their sperm and eggs would be incompatible with ours. Their blood and tissues would be immunologically rejected too (see Chapter 3).

One scenario for the distant future is the appearance of the technological singularity, an idea that goes back to 1847 when Richard Thornton wrote about "thinking machines," that they might "remedy all their own defects and then grind out ideas beyond the ken of mortal mind!" Other proponents of the idea of the singularity—the notion that at some point in the near to distant future, technology will change us so much that, essentially, all bets are off—include mathematicians Alan Turing (in 1951), Stan Ulam (1958), I. J. Good (1965), Vernor Vinge (1983), and most recently Ray Kurzweil.

The singularity is the subject of many books. There is the Singularity University (where I am a lecturer) located at NASA Moffett field in California (where else?), and even a documentary film, *Transcendent Man* (2009), and a science fiction film, *Transcendence,* in which a brain implant allows you to control a computer network. The argument goes basically that many of our key technologies, in particular computing and nanotechnology, are improving exponentially and synergizing, and as a consequence the rate of change is increasing. At some point computers may become artificially intelligent and then smarter than humans, being able to recode their own code. This seems like a perfect recipe for an autocatalytic explosion, but unlike a literal explosion there is no dilution at the end of it, just unthinkable levels of smartness.

The fifth question for vitalism (posed way back in Chapter 1) is whether consciousness (or a mind) can be engineered synthetically. This is often framed as a method of mapping a human mind into a silicon-based computer, because of the perceived potential speed of thinking, sharing, and backing up mind states. Well before then, we will continue to engineer biological minds (with words, drugs, computer interfaces, and genetics). If we freeze and thaw a midge larva, it survives. If we slice its brain in half and realign the two pieces perfectly, will it remember its previous thoughts? (Midge-level thoughts like "Gee. Is winter early this year?") Can we make variations on this brain and see the effects perhaps on a large robotic scale? (Midges are cheap.) We genomically engineer the midge to help or engineer larger animals to be more midgelike in surviving brain freeze.

Some of us might be uncomfortable extrapolating further measured exponential curves (e.g., those in Figure 7.1) or feel that theoretical limits to improvement, such as minimum size of circuits set by atoms, thermodynamics, or quantum uncertainty, may kick in soon, and impede progress. But an alternative critique accepts that we will soon be able to compute at currently inconceivable (but still finite) speeds, at which point we won't find much to compute. It may turn out that many of the most interesting things to model, to number-crunch, and to think about, are already computed at high speed by nature—for example, the many-body problem in physics for predicting trajectories of complex collections of particles, or the folding of polymers into developing humans. Analytic intelligence, where we merely observe and predict, gives way to synthetic intelligence, where we predict the future by changing it. It may turn out that well before we hit limits of computing, we will discover that we can achieve most of our goals that don't defy laws of physics without radical change from whatever passes for human nature at the time—possibly even recognizable to current humans—probably no eugenics (government control of genetic inheritance) but heavily laden with W-genics (you-eu-genics, individual control over their own body genetics) and euphenics (changing traits by changing environments, drugs, and devices).

Yet another viewpoint emerges from a play on the word "singular." As I have argued throughout this book, the future promises to be replete with diversity. Perhaps before we get to a singular state driven by greedy attention to the bottom line of productivity and ever faster thoughts, we might redefine our goals in terms of multiplicity. The counterpoint to singularity is multiarity. The alternative to fast food is the slow food movement. Perhaps we will find ways to construct and study multiverses—multiple parallel universes. We can finally not just ask "What if?" but get answers by means of diverse experiments. And we may see advantages of exploring this diversity at some intrinsic physical rate rather than following an ever accelerating exponential rate curve as championed in the singularity view.

ße   ße   ße

As a modest step toward these possible futures, we need to get at least some of our genomes and cultures off of this planet or trillions of person-years of work will be lost. We cannot assume that there is anyone at the other end to receive our radio broadcasts—or to replace us if we die. We probably won't wait until computer intelligence is as compact and flexible as biointelligence is already. The only working nanotechnology right now is bionanotechnology, and we cannot confidently predict when this will change. So we need to shoot our SCHPON (sulfur, carbon, hydrogen, phosphorus, oxygen, and nitrogen, "spawn") into the void. Probably few will survive. The term panspermia, from the Greek for "all seed" (slightly more gender neutral than it first sounds), refers to the theory that life on earth emerged from cells sent from space. Panspermia doesn't really solve the origin of life problem, but could nevertheless be part of our history eventually. In other words, we will be seeding outer space with ourselves or our descendents.

This effort to colonize the universe will benefit from engineering radiation resistance, low gravity resistance, and other such properties into our genome. If we were to convert all carbon currently in biomass into humans plus renewable food, then we could have 1,000 times as many humans as we have now (or an equivalent number of cells to send out). Adding the whole atmosphere doesn't change this much since carbon (as CO_2) is only 0.04 percent of the atmosphere. Mining the earth's crust could give another factor of 10,000 (probably limited by nitrogen), or 10^{17} humans. The total solar energy of 10^{17} watts striking earth could conceivably support 10^{14} humans at 100 W each (with up to 90 percent left over). This gives us a population density of 100,000 per sq km of total land, thirty times the current maximally dense city (Mumbai) at 30,000 per sq km. The remaining factor of 1,000 in excess biomass, which while unsustainable on earth, might be compatible with moving such mass to escape orbit at 3 GJ per 50 kg (person equivalent) over a roughly 1,000-year period (assuming a biomass doubling time of one year, rather than the current 40 year human population doubling time)—or even less time if solar energy outside our planet were harnessed.

These are not recommendations, since we imagine uncertainties and unintended consequences galore, but these numbers do provide a crucial perspective on future prospects and possibilities. As we go forward, survival reasoning may discover the optimal balance between the high diversity and intelligence of a large population versus the challenges of sustainability.

But as we go into some combination of outer and inner space, ourselves a combination of carbon and silicon-based life, we still have questions pertaining to the manner in which we will pursue our explorations. Will we be a well-stirred homogenous "optimal" monoculture, or will we be a cacophonous anarchy of self-experiments—or something in between? Why and how will we teach our robotic or H+ descendants about emotions or morality?

Which brings us back to the question, What *should* we do? What should *they* do? Hopefully this story gives you more than broad strokes and hazy crystal ball gazing—but instead lays out a recipe (a genome) for a bold recoding of nature that emphasizes diversity and safety.

Afterword

૪௨

In the year and a half since *Regenesis* was published, synthetic biology has gained momentum, both in the scientific community and in the media. It is heartening to see a growing number of scientists and policymakers exploring the possibilities of this fertile field, and what follows is an update describing a few key developments in synthetic biology since the publication of the first edition.

An Encoded Book

On October 4, 2012, just two days after this book was published, I was a guest on the *Colbert Report,* and Stephen Colbert tried to eat 20 million copies of it. Admittedly, I was not entirely blameless—I had brought as a prop a strip of paper with a tiny dot of DNA on it, with millions of copies of *Regenesis* encoded in its strands. Happily, I managed to lunge across the table and stop Colbert from ingesting our hard work.

Encoding this book in DNA has led to more than good television and increased book sales, however. After my appearance on Colbert (and the publication of a paper in *Science* about the process), several companies and museums interested in archiving contacted me about the future of digital storage. One has even funded a scaled-up version of the process I described in "Notes: On Encoding This Book into DNA." The goal of this next phase is to bring down the cost another ten-thousand-fold and to encode not only text, but several short videos in DNA.

The BRAIN Initiative

The Epilogue of this book (sections 3 and 7) discusses our interest in reading the activity of our brains, but it does not mention the revolution that was brewing in our lab as we wrote it.

In September 2011, Miyoung Chun at the Kavli Foundation and others at Gatsby and Allen foundations convened a meeting of nanoscientists and neuroscientists. Of thirty-seven scientists in attendance, six of us became very excited about the potential for new technologies and, spurred on by our coauthor, Miyoung, we prepared a white paper for the US White House Office of Science and Technology Policy (OSTP). In this paper, we described disruptive technologies newly available from genomics, nanofabrication, and synthetic biology. By the end of June 2012, this paper had evolved into a peer-reviewed publication in *Neuron*. On February 12, President Barack Obama gave his State of the Union speech and announced, "Our scientists are mapping the human brain."

Just a few minutes later, Francis Collins, the head of the National Institutes of Health, tweeted, "Obama mentions the #NIH Brain Activity Map in #SOTU." That fifty-six-character explanation barely unpacked the forty-two character original, but was just enough to prompt journalist John Markoff to search for the phrase "brain activity map," which lead him straight to our *Neuron* paper. John kept digging and published a *New York Times* story on our project, though we were far from ready to announce. We scrambled to engage more neuroscientists. Six weeks later, President Obama announced the project much more clearly and with a beautiful recursive acronym, BRAIN, standing for Brain Research Advancing through Innovative Neurotechnologies. There was a $110 million budget, already aligned with similar amounts from private foundations and a European consortium focused on computer simulations of brains.

The BRAIN project's emphasis on innovative technologies was a welcome alternative to what many feared would be an inflexible scale-it-up, grind-it-out mapping project. The inspiration for the BRAIN project is the Advanced Sequencing Technology Development Project, which brought the cost per human genome down from $3 billion to $2,000 in about six years—far faster than even the speediest electronics' exponential growth curve—while improving quality.

Indeed the breakthroughs in reading and writing DNA could play a key role in the BRAIN initiative. We have published papers on the physics of the project and have concluded that we may be close to producing a "Rosetta BRAIN," in which many old and new methods are applied to one brain in a highly integrated and high-resolution whole. This would leverage fluorescent in situ sequencing (FISSEQ, Chapter 7) and synthetic biology to establish "molecular tickertapes" of neuronal activity, as well as barcodes for each cell to determine the full developmental lineage, all RNAs, and all synaptic connections. Overall, BRAIN has been a remarkably speedy gallop through a landscape normally known for its snail rides.

The Edgy Becomes Mainstream

Synthetic biology is the outsider field most likely to play a large role in the BRAIN initiative, and the Defense Advanced Research Projects Agency (DARPA) is the agency most likely to lead this (with $50 million of the $110 million per year), since it already has a significant portfolio in that field. In particular, Alicia Jackson and Arati Prabhakar have spearheaded Advanced Tools and Capabilities for Generalizable Platforms (ATCG), first awarded in May 2012, and Living Foundries: 1,000 Molecules with proposals submitted in September 2013.

The field of synthetic biology had a strong community before the release of *Regenesis*, and it has only continued to grow. Research in the field has greatly benefited from a ten-year National Science Foundation grant known as SynBERC (Synthetic Biology Engineering Research Center). SynBERC is a bicoastal group with labs at UC-Berkeley (Chris Anderson, Adam Arkin, Jay Keasling, and Susan Marqusee), Harvard (Pam Silver and I), UC-SF (Tanja Kortemme and Wendell Lim), MIT (Natalie Kuldell, Ken Oye, Kris Prather, Chris Voigt, and Ron Weiss), and Stanford (Drew Endy). We meet at least twice a year, in the fall at MIT and in the spring at UC-Berkeley. Both the SynBERC and DARPA Living Foundries communities work with private companies in order to illuminate practical applications for synthetic biology and to enhance possible sustainability of the vision. SynBERC has a particularly effective "human practices" component, which includes topics like bioethics, safety, intellectual property, and communication with the public.

The previously super-edgy PersonalGenomes.org (PGP, see Chapter 9) has become mainstream in the past year. The National Institute of Standards and Technology (NIST) and the Food and Drug Administration (FDA) have formed a historic collaboration called Genomeinabottle.org, which will divide a large pot of standard human genomes equally into thousands of identical bottles for use in developing new diagnostics, new instruments, and new services. After an extensive search, researchers concluded that only PGP samples were suitably consented. The National Center for Biotechnology Information at the National Library of Medicine also chose PGP to be the only fully open-access dataset for human genomic, environmental, and trait data. The ENCODE Project (Encyclopedia of DNA Elements) has so far been dependent on nonideally consented cells (like HeLa cells, featured in *The Immortal Life of Henrietta Lacks,* by Rebecca Skloot), and PGP has potential as a much needed alternative. PGP is also becoming the go-to for standard cells in making custom human organoids and for human synthetic genomics to check out putative casual mutant alleles—for example, the work being done at the Center for Causal Variants at the National Human Genome Research Institute Center of Excellence in Genomic Science. PGP sites are opening up internationally, most notably PGP-Canada in December 2012 and PGP-UK in November 2013, with about a dozen more in the works.

The CRISPR Craze

Ever since the start of the genome project in December 1984, critics have emphasized the need to avoid studying junk DNA, typically defined as "repetitive and/or nonconserved DNA." But at this point it appears far easier and more informative to sequence whole genomes than their parts, and junk DNA is turning into gold.

In 1987 researchers first noticed some of this DNA in *E .coli,* and much later it was named CRISPR—clustered regularly interspaced short palindromic repeats. For decades, CRISPR was important to bacterial bioinformatics gurus, but few others took notice. Then in August 2012, Jennifer Doudna (UC-Berkeley), Emmanuelle Charpentier (Umeå University), and their colleagues published a biochemical analysis of a small part of the CRISPR story, the ninth associated protein, called Cas9, as found in the fairly obscure bacterium *Streptococcus pyogenes.* They discovered that two Cas9 RNAs (or a fusion of the two) recognize the genome based on twenty base pairs and cleave it. Although Cas9 was from a bacterium, it seemed like it might lead to a way to move multiplex genome engineering (MAGE, Chapters 3 and 6) beyond *E .coli.* This seemingly small step immediately inspired our lab as well as our collaborators at Harvard and MIT (Keith Joung, David Liu, and Feng Zhang), and within five months we began to test and publish our ideas on developing Cas9 for human genome engineering and therapies. Jennifer and the four of us from Boston soon created a startup company, Editas, to carefully test CRISPR for various human diseases. Liz Pennisi at *Science* dubbed all this new activity "the CRISPR craze." Prashant Mali and Luhan Yang, both postdoctoral fellows, helped initiate this craze at Harvard, and they continue to find new ways to gain specificity and efficiency of delivery. They have formed a second CRISPR company, Egenesis, which focuses on organ transplantation.

Revive and Restore: Mammoths

The cover of this paperback edition of *Regenesis* symbolizes the goal of restoring life to extinct animals, in particular, creating something resembling the woolly mammoth based on DNA analyzed from bones (or other body parts), synthesized chemically, and then combined with modern Asian elephant stem cells. In response to Chapter 6 of *Regenesis,* Jerry Coyne (University of Chicago) and John Hawks (University of Wisconsin) stated, respectively, that "We are a long way from taking DNA information and making a living cell from it" and "we are nowhere close to putting that

DNA onto chromosomes in the proper order . . . I doubt that we'll have this ability within the next 50 years, if that."

CRISPR has already made those statements seem very shortsighted. We can now easily place new segments of chromosomes in the proper order. For example, in human stem cells we've replaced the whole human Thy1 gene with the mouse equivalent. Just as we helped develop affordable sequencing in six years rather than six decades (noting that even six decades seemed like an aggressively optimistic exponential prediction of Moore's law), CRISPR allowed us to contract the above fifty-year prediction down to a single year. (To be fair, this was somewhat feasible even at the time of Jerry's prediction.)

Moreover, in the past year a wonderful community has come together to consider the myriad implications of using ancient and synthetic DNA to assist endangered species, de-extinctions, and resurrections. Ryan Phelan, Stewart Brand, and I hosted meetings on these topics at Harvard and at the National Geographic headquarters in Washington, DC, followed by a public meeting there (with videos online) cohosted by TEDx. The Long Now Foundation's popular Revive and Restore website posts lists of candidate species and resources.

Neanderthal News

As mentioned above, my lab does a small amount of work on synthetic biology of ancient animal DNA to test interesting hypotheses and possibly help endangered keystone species and ecosystems. In 2007, a PBS documentary maker asked me whether such approaches could be applied to ancient human DNA. Similar questions were asked by Nicholas Wade at the *New York Times* in 2009 and by Ed Regis when he interviewed me in 2010 for his *Discover* magazine article, "The Picasso of DNA." In that piece, he opined that if anyone made a Neanderthal genome in a human cell using methods from my lab, then getting the full being would require implantation "perhaps into an adventurous human female." These were Ed's words, not mine, and Ed repeated the phrase in *Regenesis* two years later. Those few comments from 2007 to 2012 provoked very little public response.

That changed dramatically in January 2013, when Spiegel Online published a German synopsis of an interview with me, which touched on this same topic and was basically accurate. The full interview was conducted in English, digitally recorded, and published a few days later nearly verbatim. (The publication apologized later for oddly inserting the word "hell," which I don't typically use.) But in those few days, between publishing the German translation and the original English, the *MIT Technology Review* had already mutated Ed's phrase into a "playful" headline: "Wanted: Surrogate for Neanderthal Baby." The misunderstanding was blamed on the German translation and on the British tabloids, but this was not really a mistranslation. Rather, a tech-savvy journalist (Susan Young) had assumed that her humorous headline would not be taken literally. Regardless, this secondhand story was picked up by many thirdhand news agencies, spawning headlines like "Harvard Professor Seeks Mother for Cloned Cave Baby" and virally spreading all over the world. Though no research team is anywhere close to needing surrogate mothers, the story captured the blogosphere's imagination. Hundreds of women sent enthusiastic emails and voice mails generously volunteering for the task. I was deeply moved by this.

Less heartening was the fact that any of the downstream bloggers could have verified this claim by checking the online *Spiegel* interview or asking me, but almost no one did. Indeed, once the *Boston Globe*, the *Boston Herald*, and *Forbes* came directly to me, the story was straightened out within a day or so (although I still get an occasional offer of surrogacy a year later). Carolyn Johnson at the *Globe* quoted me as saying, "If somebody had said that some football team had thrown a 200-yard pass, everyone would have laughed and said, 'April fools!' and something got distorted. But this is a much longer pass than a 200-yard pass." People believed it unquestioningly. After such clarifications, most comments were supportive. I work hard to keep the focus on the mundane aspects of important scientific research, and do not bring up irrelevant, sensational topics in interviews—indeed I try to deflect them. Nevertheless, when journalists insist, as they often do, I choose to respect their assessment of their readers' needs and try to use the opportunity to give some perspective on both the sensational and mundane elements of scientific research.

Teleportation

The idea of scanning a 3D object in one city, sending it via Internet, and then printing a 3D model dates back at least to 1984. This vision is still some way off, as 3D printers are challenging for mixed materials and rarely used for actual consumer goods. The resolution used these days is usually around 0.1 mm and the time taken to print typically goes up with the cube of the resolution improvement.

Because of these limitations, the idea of extending this process to atomically precise objects might benefit from new ways of thinking. The first example of synthesizing a precise genome from an electronic form was the Hepatitis C virus in 2000 (as mentioned in Chapter 3). Craig Venter's recent book, *Life at the Speed of Light*, notes that reading, sending, and writing DNA is getting easier and hence deserves the name "teleportation." The key critiques of this idea are the following: (1) This new meaning for the word teleportation falls short of the usual (sci-fi) expectations; (2) This new meaning describes something that we do already; (3) Sending only DNA information fails to cover important biological features, like epigenetics and memories; (4) The process of synthesizing whole new genomes is expensive; (5) Getting new genomes to function is, so far, limited to closely related cells; (6) The practical applications (even in the future) are not yet compelling.

One application of this type of teleportation is examining possible life on alien lands. Using this technology, we could study a life form's DNA on earth without contaminating the environment of the other planet. Indeed, maintaining the perfect sterility of landers and rovers has been a high priority for years. Chris Carr, Gary Ruvkun, Maria Zuber, and I have been working with NASA on life detection since 2006. We call our project SETG (Search for Extraterrestrial Genomes), a variation on the acronym SETI (Search for Extraterrestrial Intelligence). One key challenge has been to acquire a small and efficient sequencing device on the order of 1 kg mass and 6 watts peak power consumption. These specs were way out of range for the devices available in 2006. But today we have some emerging options. Three companies with which I have worked, Genapsys, Oxford Nanopore, and Genia, are developing palm-size instruments based on electronic detection of DNA single nucleotides in highly parallel reactions on chips. These are just now being tested in outside labs.

The key challenge mentioned in (5) above, getting radical new (or ancient or extraterrestrial) genomes to function, made a big leap forward with three papers in *Science* from our group, all published in October 2013. The process of genome editing and code changing described in Chapter 5 has now emerged from its long incubation.

The Future Beyond

One year out, we have no major amendments for the epigenetic epilogue. We do hope that you, dear reader, equipped with this handy paperback version of *Regenesis*, will join the growing wave of citizen scientists— skeptical yet energetic. Having read this book, you will not naively believe tabloids promising teleportation into the brain of a Neanderthal on Mars, but you will be equipped to actively engage in tough technical-social questions, such as: Should we now have "genomes for all," the carrier testing for thousands of highly actionable human genetic diseases? Soon, if not already, you will be studying your own microbiome and immunome, and even changing them, whether connected with routine health care or on your own. And I can only hope some of you will be conspiring with your creative filmmaking friends about making an epic possibly titled *Regenesis: The Movie*—or more likely, *My Ome: The YouTube*.

ACKNOWLEDGMENTS

GEORGE CHURCH

Regenesis would have been hard for Ed Regis or me to have written separately, so we've tried to share the work, credit, and blame. We hope to develop sequels, movies, smartphone apps, and action figures. So if we've inadvertently omitted someone you admire and you'd like to secure a place in history for that someone and/or yourself, please let us know, and we'll try to re-reroute the rivers Alpheus and Peneus to include your nominees in our next labors.

I'd like to acknowledge my coworkers, family, and associates, especially Ting Wu and John Aach, who have helped in innumerable ways since 1978 and 1995, respectively, including (but not limited to) safety, public outreach, ethics, and brilliant theoretical and experimental insights. My 1984 PhD thesis acknowledged 258 people (arep.med.harvard.edu/gmc/ackk.html), and that list remains relevant—especially my high school, college, grad school, and postdoc mentors, Crayton Bedford, Sung-Hou Kim, Wally Gilbert, and Gail Martin. I'm very grateful to roughly four hundred journalists who, since 1996, have helped me rise slightly above my feral dream-speak to something closer to intelligible (http://arep.med.harvard.edu/gmc/news.html). One of these, of course, was Ed. A total of 1,274 coauthors on previous publications have greatly shaped the knowledge base from which I have drawn in these pages (arep.med.harvard.edu /gmc_pub.html). Sri Kosuri graciously guided me through the tricky parts of experiments I did to write and read this book in DNA form.

PGP staff and volunteers have been inspiring and patient. The core of this book concerns the synthetic biology community and I am indebted to my entrepreneurial BioFab colleagues (Joe Jacobson, Tom Knight, Drew Endy, and Jay Keasling) and the Wyss Institute for Biologically Inspired Engineering (especially professors Don Ingber, Jim Collins, Pam Silver, Peng Yin, and William Shih). Showing great courage and trust were private donors and program directors for granting agencies NSF, DARPA, DOE, NIH, and Google, who supported this work when it was raw, speculative, and unpopular. John Brockman, definitively the most amazing literary champion of popular, edgy, intellectually provocative works, was hugely responsible for this, my first, book. When my draft was rejected multiple times by publishers, John did not abandon me but doubled his efforts. When John mentioned the option of a "ghost writer," as employed by some of my colleagues, I realized that writing collaboratively resonated with my nature and would result in a much better book. But I would only be comfortable with a living and fully acknowledged coauthor (not a ghost). Ed was a terrific choice, with numerous popular science books under his belt already, on topics ranging from Einstein's office to Mambo chickens. This time, with Ed's help, we had no problem getting publisher enthusiasm. At Basic Books, TJ Kelleher and his staff were great. TJ confidently overruled my concerns over the cover art and book title. He noted that despite his jaded feeling that "at a certain point one becomes hard to impress. My initial reaction was scarcely mental—it was just an emphatic, emotional response; my skin exploded in goose bumps. I was speechless, and when I finally got something out I asked my art director to make me a poster of it." Since you, dear reader, made it past the cover and are reading this page now, perhaps you had a reaction similar to TJ's and, if so, I hope that it applies to the rest of your journey through this book.

ED REGIS

I first met George Church on Tuesday, May 19, 2009, when I interviewed him in his office at the Harvard Medical School for a profile to be published in *Discover* magazine. (It ran in the March 2010 issue as "The Picasso of DNA.") But it was a meeting that almost didn't happen. Church has one of the most complicated and jam-packed schedules known to humankind (kept by not one but two administrative assistants), including meetings with as many as eight individuals (or groups) on any given day, not to mention the further tasks of teaching classes, planning and overseeing lab experiments, writing scientific papers and reports, attending conferences, and so on. (The interested reader may view Church's current schedule at http://arep.med.harvard.edu/labmeeti.html#tour.)

The day before, when I checked in at my Boston hotel and read my email, a note from Church appeared informing me that since he was in Philadelphia attending a scientific conference, he would have to reschedule my interview on one of a series of dates he proposed. Fortunately, I had left an "away from keys" message on my email account. Seeing the message, Church wrote an immediate follow-up note telling me that he would fly back to Boston that night and meet me the next morning as planned. Which he did.

Both Church and I have the same literary agent, John Brockman, and it was Brockman who suggested that we collaborate on a book. The result is before the reader, and so my first thanks go to Brockman and to his partner, Katinka Matson.

Being a collaborating writer was a first-time experience for me, and I had often wondered how those who did it managed the task. Now I know . . . or at least I know how George and I did it—through a relatively simple division of labor plus repeated efforts to correct and improve each other's work. In general, a section that looks as if it must have been written by a scientist was probably written mainly by George. By contrast, sections that appear to be the product of a mere science writer are most likely by myself. Many thanks to George Church for being such a wildly inventive, deeply thoughtful, and extremely cooperative cowriter. This was lots of fun, and I would do it again in a minute.

I don't necessarily agree with all viewpoints expressed in these pages. Although it's by both of us, the book is written in George's voice, and it expresses his personal vision.

My previous books were edited by generalists. Our editor on this book, Thomas J. Kelleher, is a specialist. In John Brockman's words, he's "a real science editor." His perceptive notes on prior versions of the text materially improved its organization, presentation, and readability, and for this I am greatly indebted to him.

In addition, the entire text was read by three experts: Harold Morowitz (Robinson Professor in Biology and Natural Philosophy, George Mason University), Claire Fraser-Liggett (professor of medicine and director, Institute for Genome Sciences, University of Maryland School of Medicine), and Anthony C. Forster, MD, PhD (University Chair in Chemical Biology, Uppsala University). I am grateful to all three for spotting errors and omissions, and for requesting clarifications and corrections. Any errors that remain are of course the sole responsibility of the authors.

Research on this project was supported in part by a grant from the Alfred P. Sloan Foundation. I would like to thank Doron Weber, vice president, programs, for championing my cause once again.

Additional thanks go to my wife, Pamela Regis, and to Chris Anderson, Misha Angrist, Jose Folch, David Donnell, Eri Gentry, Gary C. Hudson, Kevin Kelly, Meagan Lizarazo, Nils Lonberg, J. P. de Magalhaes, Jay Valdes, and Rodney E. Willoughby Jr., MD.

BOTH

Grateful acknowledgment is made of a short (14 word) quote under the doctrine of fair use (the quote appears 20,000 times on the Internet as well): James Joyce, *A Portrait of the Artist as a Young Man* (1916), and eight words written by Richard Feynman.

SELECTED REFERENCES

PROLOGUE

Carlson, Robert H. *Biology Is Technology*. Cambridge: Harvard University Press, 2010.

Chang, Timothy Z. "Kinetics of Wastewater Treatment and Electricity Production in a Novel Microbial Fuel Cell." *Journal of the U.S. SJWP*, 2008, 16–31.

Evans, Jon. "Bioplastics Get Growing." *Plastics Engineering*, February 2010, 14–19.

Folch, J., et al. "First Birth of an Animal from an Extinct Subspecies (*Capra pyrenaica pyrenaica*) by Cloning." *Theriogenology*, 2009. doi:10.1016/j.theriogenology.2008.11.005.

Kolata, Gina. *Clone: The Road to Dolly and the Path Ahead*. New York: William Morrow, 1998.

Liu, Hong, et al. "Production of Electricity During Wastewater Treatment Using a Single Chamber Microbial Fuel Cell." *Environmental Science and Technology* 38, no. 7 (2004): 2281–2285.

Ziegler, Julie. "Metabolix Defies Skeptics with Plastic from Plants." Bloomberg.com, May 2009.

Web

ir.metabolix.com/releasedetail.cfm?ReleaseID=239056
mirelplastics.com
dupont.com/Sorona_Consumer
scarabgenomics.com
inhabitat.com/grow-your-own-treehouse

CHAPTER 1

Amitābha Sūtra
Genesis 22:17
Brock, William H. *The Norton History of Chemistry*. New York: Norton, 1992.

Web

en.wikipedia.org/wiki/Hadean

"Transplant Jaw Made by 3D Printer Claimed As First." BBC News, February 6, 2012. bbc.co.uk/news/technology-16907104

CHAPTER 2

Burgess, Jeremy, et al. *Under the Microscope*. Cambridge: Cambridge University Press, 1990.

Cook-Deegan, Robert. *The Gene Wars*. New York: Norton, 1994.

Duve, Christian de. *A Guided Tour of the Living Cell*. 2 vols. New York: Scientific American Books, 1984.

Forster, Anthony C., and George M. Church. "Towards Synthesis of a Minimal Cell." *Molecular Systems Biology* 2 (2006), 45.

Gibson, Daniel G., et al. "Creation of a Bacterial Cell Controlled by a Chemically Synthesized Genome." *Science Express*, May 20, 2010, 10.

Gibson, Daniel G., et al. "Complete Chemical Synthesis, Assembly, and Cloning of a *Mycoplasma genitalium* Genome." *Science*, February 29, 2008, 1215–1220.

Glass, John L., et al. "Essential Genes of a Minimal Bacterium." *PNAS*, January 10, 2006, 425–430.

Goodsell, David S. *The Machinery of Life*. New York: Springer-Verlag, 1993.

Jeon, K. W., et al. "Reassembly of Living Cells from Dissociated Components." *Science*, March 20, 1970, 1626–1627.

Lartigue, Carole. "Creating Bacterial Strains from Genomes That Have Been Cloned and Engineered in Yeast." *Science*, September 26, 2009, 1693–1696.

Nagel, Thomas. "What Is It Like to Be a Bat?" *Philosophical Review*, October 1974, 435–450. organizations.utep.edu/Portals/1475/nagel_bat.pdf.

Russo, Eugene. "The Birth of Biotechnology." *Nature*, January 23, 2003, 456–457.

Web

en.wikipedia.org/wiki/Archean

CHAPTER 3

Blight, Keril J., et al. "Efficient Initiation of HCV RNA Replication in Cell Culture." *Science*, December 8, 2000, 1972–1974.

Cello, Jeronimo, et al. "Chemical Synthesis of Poliovirus cDNA: Generation of Infectious Virus in the Absence of Natural Template." *Science*, August 9, 2002, 1016–1018.

deMagalhaes, J. P., et al. "A Proposal to Sequence Genomes of Unique Interest for Research on Aging." *Journal of Gerontology* 62A (2007): 583–584.

Douglas, S. M., et al. "Rapid Prototyping of Three-Dimensional DNA-Origami Shapes with caDNAno." *Nucleic Acids Res* 2009. doi:10.1093/nar/gkp436. See also caDNAno.org.

Douglas, S. M., I. Bachelet, and G. M. Church. "A Logic-Gated Nanorobot for Targeted Transport of Molecular Payloads." *Science* 335 (2012): 831–834.

Isaacs, Farren J. et al. "Precise Manipulation of Chromosomes In Vivo Enables Genome-Wide Codon Replacement." *Science,* July 15, 2011, 348–353.

Schrödinger, Erwin. *What Is Life? The Physical Aspect of a Living Cell.* Cambridge: Cambridge University Press, 1945.

Seeman, Nadrian. "Nucleic Acid Junctions and Lattices." *Journal of Theoretical Biology* 99 (1982): 237–247.

Seluanov, Andrei, et al. "Hypersensitivity to Contact Inhibition Provides a Clue to Cancer Resistance of Naked Mole-Rat." *PNAS* 106, no. 46 (2009): 19352–19357. doi:10.1073/pnas.0905252106.

Smith, Hamilton O., et al. "Generating a Synthetic Genome by Whole Genome Assembly: phiX174 Bacteriophage from Synthetic Oligonucleotides." *PNAS,* December 23, 2003, 15440–15445.

Van Hemert, M. J., et al. "SARS-Coronavirus Replication/Transcription Complexes Are Membrane-Protected and Need a Host Factor for Activity." *Vitro.PLoS Pathog* 4, no. 5 (2008), e1000054. doi:10.1371/journal.ppat .1000054.

Wang, Harris H., et al. "Programming Cells by Multiplex Genome Engineering and Accelerated Evolution." *Nature,* August 13, 2009, 894–897.

Yu, C., et al. "RNA Sequencing Reveals Differential Expression of Mitochondrial and Oxidation Reduction Genes in the Long-Lived Naked Mole-Rat (*Heterocephalus glaber*) When Compared to Mice." *PLoS ONE* 2011, 6:e26729.

CHAPTER 4

Broad, William J. "Tracing Oil Reserves to Their Tiny Origins." *New York Times,* August 2, 2010.

Demerjian, Dave. "Aviation Biofuels: More Hype Than Hope?" wired.com/autopia/2009/01/aviation-bio-fu.

Glasby, Geoffrey P. "Abiogenic Origin of Hydrocarbons: An Historical Overview." *Resource Geology* 56 (2006): 85–98.

Gold, Thomas. "The Origin of Methane (and Oil) in the Crust of the Earth." USGS Professional Paper 1570. *The Future of Energy Gases,* 1993.

Ingram, L. O., et al. "Genetic Engineering of Ethanol Production in *Escherichia coli.*" *Applied and Environmental Microbiology* 53 (1987): 2420–2425.

Knothe, Gerhard. "Historical Perspectives on Vegetable Oil-Based Diesel Fuels." *Inform* 12 (2001): 1103–1107.

Naylor, Rosamond L., et al. "The Ripple Effect: Biofuels, Food Security, and the Environment." *Environment* 49 (2007): 31–43.

Page, Lewis. theregister.co.uk/2008/12/30/oily_kiwi_nut_jet_juice.
theregister.co.uk/2008/02/25/virgin_747_coconut_yes_algae_no
theregister.co.uk/2009/01/09/algae_airliner_test_success

Robertson, D. E., et al. "A New Dawn for Industrial Photosynthesis." *Photosynth Res.* March 2011, 269–277.

Rosenthal, Elizabeth. "Rush to Use Crops as Fuel Raises Food Prices and Hunger Fears." *New York Times*, April 6, 2011.

Schirmer, A., et al. "Microbial Biosynthesis of Alkanes." *Science,* July 2010, 559–562.

Steen, Eric J., et al. "Microbial Production of Fatty-Acid-Derived Fuels and Chemicals from Plant Biomass." *Nature,* January 28, 2010, 559–563. doi:10.1038/nature08721.

Voosen, Paul. "As Algae Bloom Fades, Photosynthesis Hopes Still Shine." *New York Times*, March 29, 2011.

Web

aef.org.uk/uploads/Bio_fuelled_or_bio_fooled_article__2_.pdf
en.wikipedia.org/wiki/Carboniferous
econlib.org/library/Bastiat/basEss1.html
getg.com/engineOil.php
jouleunlimited.com
ls9.com
marinecorpstimes.com/news/2010/10/marine-push-for-alternative-fuels-102510
sapphireenergy.com
ucmp.berkeley.edu/carboniferous/carboniferous.php
walmart.com/ip/G-Oil-5W-30-Bio-Synthetic-Motor-Oil-5qt/16928048

CHAPTER 5

Anderson, J. Christopher, et al. "Environmentally Controlled Invasion of Cancer Cells by Engineered Bacteria." *Journal of Molecular Biology* 355 (2006): 619–627.

Clark, William R. *At War Within: The Double-Edged Sword of Immunity.* New York: Oxford University Press, 1995.

Comins, Neil F. *The Hazards of Space Travel.* New York: Villard, 2007.

Gladyshev, Eugene, and Matthew Meselson. "Extreme Resistance of Bdelloid Rotifers to Ionizing Radiation." *PNAS*, April 1, 2008, 5139–5144.

Hu, William T., et al. "Long-Term Follow-up After Treatment of Rabies with Induction of Coma." *New England Journal of Medicine*, August 30, 2007, 945–946.

Isaacs, Farren J. et al. "Precise Manipulation of Chromosomes In Vivo Enables Genome-Wide Codon Replacement." *Science,* July 15, 2011, 348–353.

Johnson, Mark. "Rabies Survivor Jeanna Giese Graduates from College." json-line.com/news/wisconsin/121479779.htmlalso jeannagiese.com.

Lonberg, Nils. "Human Antibodies from Transgenic Animals." *Nature Biotechnology,* September 2005, 1117–1125.

———. Email to Ed Regis, July 7, 2010.

Willoughby, Rodney Jr. Email to Ed Regis, August 31, 2010.

Willoughby, Rodney E. Jr., et al. "Survival After Treatment of Rabies with Induction of Coma." *New England Journal of Medicine,* June 16, 2005, 2508–2514.

CHAPTER 6

Briggs, Robert, and Thomas J. King. "Transplantation of Living Nuclei from Blastula Cells into Enucleated Frogs' Eggs." *PNAS* 38, 1952: 455–463.

Burbano, Hernán A., et al. "Targeted Investigation of the Neandertal Genome by Array-Based Sequence Capture." *Science,* May 7, 2010, 723–725.

Folch, J. Email to Ed Regis, August 20, 2010.

Folch, J., et al. "First Birth of an Animal from an Extinct Subspecies (*Capra pyrenaica pyrenaica*) by Cloning." *Theriogenology,* 2009. doi:10.1016/j.theriogenology.2008.11.005.

Gray, Richard, and Roger Dobson. "Extinct Ibex Is Resurrected by Cloning." *The Telegraph,* 31 January 2009.

Green, Richard E., et al. "Analysis of One Million Base Pairs of Neanderthal DNA." *Nature,* November 16, 2006, 330–336.

———. "A Draft Sequence of the Neandertal Genome." *Science,* May 7, 2010, 710–722.

Kolata, Gina. *Clone: The Road to Dolly and the Path Ahead.* New York: William Morrow, 1998.

Kolbert, Elizabeth. "Sleeping with the Enemy." *New Yorker,* August 15, 2011, 64–75.

Nicholls, Henry. "Darwin 200: Let's Make a Mammoth." *Nature* 456 (2008): 310–314.

Olson, Steve. "Neanderthal Man." *Smithsonian,* October 2006.

Pääbo, S. "Molecular Cloning of Ancient Egyptian Mummy DNA." *Nature* 314 (1985): 844–845.

Stone, Richard. *Mammoth: The Resurrection of an Ice Age Giant.* Cambridge, MA: Perseus, 2001.

Wade, Nicholas. "Regenerating a Mammoth, for $10 Million." *New York Times,* November 19, 2008.

Wilmut, I., et al. "Viable Offspring Derived from Fetal and Adult Mammalian Cells." *Nature* 385 (1997): 810–813.

Zimov, Sergey A. "Pleistocene Park: Return of the Mammoth's Ecosystem." *Science* 308 (2005): 796–798.

Zorich, Zack. "Should We Clone Neanderthals?" *Archaeology,* March-April 2010.

Web

en.wikipedia.org/wiki/Pleistocene
en.wikipedia.org/wiki/Neanderthal

CHAPTER 7

Baker, D., et al. "Engineering Life: Building a FAB for Biology." *Scientific American,* June 2006, 44–51.

Blanchard A., R. J. Kaiser, and L. E. Hood. "High Density Oligonucleotide Arrays." *Biosensors and Bioelectronics* 11 (1996): 687–690.

Carroll, Lewis. *Through the Looking-Glass, and What Alice Found There.* 1871.

Shreeve, James. *The Genome War: How Craig Venter Tried to Capture the Code of Life and Save the World.* New York: Knopf, 2004.

Staden, Roger. "A Strategy of DNA Sequencing Employing Computer Programs." *Nucleic Acids Res.* 6 (1979): 2601–2610.

Sulston, John, and Georgina Ferry. *The Common Thread.* Washington, DC: Joseph Henry, 2002.

Web

en.wikipedia.org/wiki/Neolithic

CHAPTER 8

Campbell, A. Malcolm. "Meeting Report: Synthetic Biology Jamboree for Undergraduates." *Cell Biology Education* 4 (2005): 19–23.

Davis, J. "Microvenus." *Art Journal* 55, no. 1 (1996). www.jstor.org/pss/777811.

Donnell, David. Emails to Ed Regis, September 14, 2010; October 1, 2010; January 28, 2011.

Elowitz, Michael, and Stan Leibler. "A Synthetic Oscillatory Network of Transcriptional Regulators." *Nature* 403 (2000): 335–338.

Endy, Drew. "Foundations for Engineering Biology." *Nature,* November 24, 2005, 449–453.

Ferber, Dan. "Microbes Made to Order." *Science,* January 9, 2004, 158–161.

Gibbs, W. Wayt. "Art as a Form of Life." *Scientific American,* April 2001, 37–39.

Gustafsson, Claes. "For Anyone Who Ever Said There's No Such Thing as a Poetic Gene." *Nature,* April 8, 2009, 703.

Hapgood, Fred. *Up the Infinite Corridor: MIT and the Technical Imagination.* Cambridge, MA: Perseus, 1993.

Knight, Tom. Email to Ed Regis, September 10, 2010.

———. "Idempotent Vector Design for Standard Assembly of Biobricks." MIT Artificial Intelligence Laboratory, 2003. web.mit.edu/synbio/release/docs/biobricks.pdf.

Levskaya, A. "Synthetic Biology: Engineering *Escherichia coli* to See Light." *Nature,* November 24, 2005, 441–442.

Mooallem, Jon. "Do-It-Yourself Genetic Engineering." *New York Times Magazine,* February 10, 2010.

Morton, Oliver. "Life, Reinvented." *Wired* 13 (2005).

Surowiecki, James. "Turn of the Century." *Wired* 10 (2002).

Web

2010.igem.org/Team:The_Citadel-Charleston/TeamPage

2010.igem.org/Main_Page

2010.igem.org/Team:Hong_Kong-CUHK

biobricks.organdpartsregistry.org

edge.org/3rd_culture/endy08/endy08_index.html

CHAPTER 9

Angrist, Misha. *Here Is a Human Being: At the Dawn of Personal Genomics.* New York: HarperCollins, 2010.

Church, George M. "Genomes for All." *Scientific American,* January 2006.

Darcé, Keith. "Study: Genetic Tests Do Little to Change Habits." *San Diego Union-Tribune,* January 17, 2011.

Hall, Stephen S. "The Genome's Dark Matter." *Technology Review,* January-February 2011.

Harmon, Amy. "My Genome, Myself: Seeking Clues in DNA." *New York Times,* November 17, 2007.

Judson, Olivia. "The Human Phenome Project." *New York Times,* June 8, 2010.

Kolata, Gina. "Staph's Trail Points to Human Susceptibilities." *New York Times,* December 15, 2010.

Lunshof, J. E., Bobe, J., Aach, J., Angrist, M., Thakuria, J. V., Vorhaus, D. B., Hoehe, M. R., Church, G. M. "Personal genomes in progress: from the human genome project to the personal genome project." *Dialogues Clin Neurosci.* 2010; 12(1):47–60.

Pinker, Steven. "My Genome, My Self." *New York Times,* January 11, 2009.

Shannon, C. E. "A Mathematical Theory of Communication." *Bell System Technical Journal* 27 (1948): 379–423, 623–656.

EPILOGUE

Bailey, Ronald. "The Case for Enhancing People." *New Atlantis,* Summer 2011.

Carlson, Robert H. *Biology Is Technology.* Cambridge: Harvard University Press, 2010.

Church, George. "Let Us Go Forth and Safely Multiply." *Nature,* November 24, 2005, 423.

———. "A Synthetic Biohazard Nonproliferation Proposal," 2004. arep.med.harvard.edu/SBP/Church_Biohazard04c.htm.

Crichton, Michael. *The Andromeda Strain.* New York: Knopf, 1969.

Donnell, David. Email to Ed Regis, October 1, 2010.

Enriquez, Juan, and Steve Gullans. *Homo Evolutis*. Ted Books, 2011.

Fleming, Diane O., and Debra Long Hunt, eds. *Biological Safety: Principles and Practices*. American Society for Microbiology, 2006.

Fukuyama, Francis. "Transhumanism." *Foreign Policy*, September 1, 2004.

"Garage Biology." *Nature*, October 7, 2010, 634.

Garfinkel, Michele S., et al. "Synthetic Genomics: Options for Governance." J. Craig Venter Institute/Center for Strategic and International Studies/MIT, October, 2007.

Gentry, Eri. Email to Ed Regis, October 28, 2010.

Ledford, Heidi. "Life Hackers." *Nature*, October 7, 2010, 650–652.

Pais, Abraham. *Inward Bound: Of Matter and Forces in the Physical World*. Oxford: Clarendon, 1986.

Parker, E. S., et al. "A Case of Unusual Autobiographical Remembering." *Neurocase*, February 2006, 35–49.

Popkin, Jim. "Authorities in Awe of Drug Runners' Jungle-Built, Kevlar-Coated Supersubs." *Wired*, April 2011.

Price, Jill. *The Woman Who Can't Forget*. New York: Free Press, 2008.

Regis, Ed. *Great Mambo Chicken and the Transhuman Condition*. Reading, MA: Addison-Wesley, 1990.

Schmidt, Markus. "Diffusion of Synthetic Biology: A Challenge to Biosafety." *Systems and Synthetic Biology* 2 (2008): 1–6. doi:10.1007/s11693–008–9018-z.

Specter, Michael. "A Life of Its Own: The Future of Synthetic Biology." *New Yorker*, September 28, 2009, 56–65.

Taleb, Nassim Nicholas. *The Black Swan: The Impact of the Highly Improbable*. New York: Random House, 2007.

Tucker, Jonathan B., and Raymond A. Zilinkas. "The Promise and Perils of Synthetic Biology." *New Atlantis*, Spring 2006, 25–45.

Wolfram, Stephen. *A New Kind of Science*. Champaign, IL: Wolfram Media, 2002.

Web

biocurious.org
biohack.sourceforge.net
diybio.org

ILLUSTRATION SOURCES

NOTES

Bancroft, C., et al. "Long-term Storage of Information in DNA." *Science* 293 (2001): 1763–1765.

Church, G. M., Y. Gao, and S. Sri Kosuri. "Next-Generation Digital Storage in DNA." Submitted 2012.

Davis, J. "Microvenus." *Art Journal* 55, no. 1 (1996). www.jstor.org/pss/777811.

———. "Romance, Supercodes, and the Milky Way DNA." *Ars Electronica* 2000.

FIGURE CREDITS

Figure 1.1 *Hands Make Anti-hands* . . . Sculpture. George M. Church (GMC) and Marie Wu.

Figure 1.2 Handedness. From *Hands Make Anti-hands*. Rasmol image from coordinates of an amino acid. umass.edu/microbio/rasmol.

Figure 1.3 Chiral crystals. en.wikipedia.org/wiki/File:Pcrystals.svg.

Figure 1.4 Three base pairs. GMC.

Figure 2.1 Liposome and protein pore. GMC, based on RCSC 7AHL.pdb coordinates and users.humboldt.edu/rpaselk/C438.S11/C438Notes/C438nLec06.htm.

Figure 3.1 Transfer RNA. GMC, combining Kim et al. "The General Structure of Transfer RNA Molecules," *PNAS* 71 (1974): 4970–4974 (Figure 2); and Xiao et al., "Structural Basis of Specific tRNA Aminoacylation by a Small In Vitro Selected Ribozyme." *Nature* 254 (2008): 358–362.

Figure 3.2 Genetic code clock. GMC, inspired by http://en.wikipedia.org/wiki/File:GeneticCode21-version-2.svg.

Figure 3.3 MAGE and CAGE. GMC, adapted from Isaacs et al., "Precise Manipulation of Chromosomes In Vivo Enables Genome-wide Codon Replacement," *Science* 333 (2011): 348–353.

Figure 3.4 DNA clock bondage. GMC, inspired by Salvador Dali's *Persistence of Memory* (1931).

Figure 3.5 DNA log pile. GMC, adapted from Shawn M. Douglas, et al., "Rapid Prototyping of 3D DNA-origami Shapes with caDNAno," *Nucleic Acids Research,* August 2009, doi:10.1093/nar/gkp436.

Figure 5.1 Antibody. en.wikipedia.org/wiki/Antibody.

Figure 5.2 Six Arg codons. See Figure 3.2.

Figure 5.3 Six Leu codons. See Figure 3.2.

Figure 5.4 The smallest viral genome. GMC, using NCBI NC_001417.

Figure 6.1 Frozen mammoth. en.wikipedia.org/wiki/File:Jeune_mammouth _IRSNB.JPG.

Figure 7.1 Exponentials. GMC; see also Carr and Church, "*Genome Engineering*," *Nature Biotechnology* 27 (2009): 1151–1162.

Figure 8.1 Biobrick assembly. GMC, inspired by partsregistry.org/Assembly: Standard_assembly.

Figure 8.2 Hello World. openwetware.org/wiki/IGEM:MIT/Sponsorship or en.wikipedia.org/wiki/File:UT_HelloWorld. Jpg from UT Austin/UCSF iGEM team.

Figure Epilogue Prohibition plot. ER adapted from Kevin Kelly, *What Technology Wants* (New York: Viking, 2010), p. 241. Used by permission.

Cover Based on *The Creation of the World,* by Eustache Le Sueur (1617–1655), Musee des Beaux-Arts, Tourcoing, France.

NOTES: ON ENCODING THIS BOOK INTO DNA

As I explain in greater detail below, to test DNA as a super-compact storage system, this book was encoded into DNA, and the resulting sequence was amplified until 40 billion copies of the DNA book had been produced. In what follows I discuss some of the legal, policy, biosafety, and other issues and opportunities pertaining to this process.

First, we'd like a compact, platform-independent file allowing images and robust to errors (mutations). Word processing formats are not really standard. PDF and many compression formats are brittle to small file errors. HTML is easily human readable and has an option for inline images (e.g., encoded in base-64 format).

Second, we consider unintended consequences; ethical, legal, social, policy, security, and safety issues; and contact diverse thought leaders.

(A) DNA could be a form of cryptography, and hence it includes legal restrictions and import/export regulations. The DNA could encode computer viruses (I embedded a mouse tracking program to symbolize a spyware threat) or the computer code could contain human viruses (see item C)—or the whole thing could go viral in a social networking sense.

(B) NIH guideline: "If the synthetic DNA segment is not expressed in vivo as a biologically active polynucleotide or polypeptide product, it is exempt from the NIH Guidelines." Even though these 156 bp long fragments are unlikely to replicate on their own or encode anything biologically active, if placed in the wild, they could get incorporated into a living organism. EPA, FDA, OSHA, USDA, and DHS guidelines might play out similarly to scenario A above.

(C) Certain apparently innocuous digital documents (e.g., images) once converted to DNA by the methods described here could result in infectious DNA molecules, so rules governing DNA surveillance (e.g., "Screening Framework Guidance for Providers of Synthetic Double-Stranded DNA" published by DHHS in the *Federal Register*, November 2009). This document's HTML code, when

279

converted to DNA form, could (but doesn't) encode parts of one or more select agents, which would have set off alarms at the facility manufacturing the DNA (unless prior explicit justification is on file). However, the current guidelines are only recommendations and apply only to sequences "longer than 200 base pairs (bps)" or "66 amino acid sequences."

(D) Incorporating DNA into our daily lives and at large scale could make us more habituated or apathetic with respect to safety and security. Yet constant exposure to cars has brought about better safety engineering (shoulder harnesses, airbags, infant carriers, etc.). Ditto for child- and tamper-resistant drug bottles, street lighting, and so on.

Third, how much does such printing cost? We print the initial DNA onto a 2-inch DNA chip for about $1,000. After sequencing to see if any sequences are underrepresented, an additional chip can be made using the redundancy of the coding to make those segments in many ways. From the combined original I made copies via polymerase chain reaction (PCR). Fifty dollars is enough to make 40 billion copies of the book, which if printed onto the first 20,000 book jackets, at 200,000 (barely visible) dots per cover (each dot containing 10 copies of the book) would be more than the sum of the top 150 printed volumes of all time, including *A Tale of Two Cities, Le Petit Prince*, Hong lou meng, the Bible, the Qur'an, *Webster's Dictionary*, Xinhua Zidian, *Boy Scout Manual, Guinness Book of World Records, Don Quixote*, and the full works of Tolkien, J.K. Rowling, Mao, Agatha Christie, and Shakespeare. The point is simply that this printing method is inexpensive and 100 million times more compact than Blu-ray disc data.

Fourth, we need to handle various issues with DNA synthesis and amplification, for example, underrepresentation of extremely high or low G+C content, inverted repeats, and runs of bases (problematic for synthesis and/or sequencing, e.g., GGGG . . .). We want to minimize missing sections of DNA, and if some are missing have the ability to order the rest. The early draft of this book that I encoded consisted of 53,418 words, 11 JPG images, and 1 Javascript program. The 644 Kbytes fit into 9 M base pairs in 91,401 DNA segments, which I synthesized at our Agilent facility and amplified using PCR. Each of these DNA segments is 96 bp long (encoding 12 bytes). The segment length is set by practical limits of oligo synthesis (Chapter 8). Each oligo begins on the left with an amplification/sequencing 22-mer primer and then a segment number L bits long. For this book, we chose L = 19 (allowing numbering up to 524,288 oligos = 6 Mbytes of text). Next is the payload 96-mer, then finally a 24-mer primer on the far right. So the full oligos are 22 + 19 + 96 + 22 = 159 bp long. The use of segment numbers means that we don't depend on overlapping sequences typical of genome assembly, a very problematic practice due to sequence repeats. In principle, the segment number could have a synthesis and/or sequencing error, but we require that the (consensus) sequence read between the primers be the correct length (115 bp) and that each 115 bit sequence be observed multiple times to be taken seriously. We can have several encodings of each segment to minimize the impact of any particular encoding that might be intrinsically hard to synthesize, amplify, or sequence. To enable multiple encodings, the zero bit can be either A or C and the one bit G or T. This can be done randomly or in a manner that minimizes problematic sequences. In the 2-bit/bp code described in Chapter 8, we observe (and underline) the unfortunate CCCCCCC four times in the example below—the inevitable consequence of the last two letters of the word OUT (01010101 01010100 = ASCII "UT"). But these same three instances of those 16 bits were encoded as aTcGaGcTcTcGaGac, cGaGaGaGaTcTaTac, aGcTcTaGaTcTaTac in the 1-bit/bp degenerate code (lowercase a or c for zero, uppercase **G** or **T** for one). DNA supercoding (1.5 bits/bp) from Joe Davis in 2000 was a step in a similar direction, enabling a degenerate base-20 encoding with triplets and exceptions; nevertheless, a 3,867 bp DNA, encoding a photo of the Milky Way, still had five instances of CCCCCCC. In 2004 his DNA manifolds pointed to a potential way to stabilize coded messages in living cells by embedding them in natural protein coding regions at 0.3-bit/bp. We chose the 1-bit/bp code for this book.

Fifth and last, we would need a convenient way to read this form of the book—a DNA reader analogous to, for example, the Kindle, Sony-PRS, Android, or iPad. This is discussed briefly in Chapter 8. Sequencing this book using Illumina technology resulted in 692 million paired 100 base pair reads for $2,000. At 91K reads per book, that means $0.26 per 1X reading of the book. No words were lost in the process. Handheld DNA readers are beginning to become available, for example, the 2012 Oxford Nanopore MinIon (Chapter 7), and milestones like this book in DNA might accelerate commercial interest. We have established a DNA Encoded Artifacts Registry (DEAR) to coordinate global use so that mixed samples and updates can be interpreted. Coordination with the Rosetta Project (of the Long Now Foundation) could help make DNA the time capsule. Rosetta 3-inch nickel disk aims at 10,000 year archiving, avoiding the pitfalls of ephemeral digital standards (and even cultures) by intuitively leading the discoverer to higher and higher magnifications and 1,500 languages. In principle, instructions for building a DNA reader could be included in one or more languages and images.

\# \# \# Begin notes and 2-bit/bp DNA encoding notes below \# \# \#
\#Decoding self-referential DNA that encodes these notes.
\#\Perl\bin\perl bp2bit.pl GMC 25-Jun-02011

```
sub b2b {return unpack("N", pack("B32", substr("0" x 32 . shift, -32)));} open
OUT,">OUT";binmode OUT;open IN,"IN"; $d{"A"}="00"; $d{"C"}="01";
$d{"G"}="10"; $d{"T"}="11"; while ($text =<IN>) { while ($text =~ s/(^.{4})//i)
```

```
{#print " $text "; $b8=chr(b2b($d{substr($1,0,1)}.$d{substr($1,1,1)}.$d{substr
($1,2,1)}.$d{substr($1,3,1)}])); print OUT "$b8";}}
```

#AGAT (00 10 00 11 #) CACA (01 00 01 00 D) AGATCACACGCCC-
GATCGTTCGCACGGCCGTGCGCTAGAACTATCGCCCGTACGC-
GAGTCCTAGCGCCCGCGCGCCCTAGCGCCCGTGCTCACGGCCGACC
GTAAGAACACACATGCAACAGAACTCACGGACGACCTCAA-
GAACGCCCGTGCGATCGTTCGCACGCCCTATAGAACTCACG-
GACGCCCTATCGCCAGAACGTGCGTTCTCACGCCCTATAGTGAATCA
AGGAGATCCTACCAACGCCCTAGCGTACCTACGAGCGGCCGTGCC-
TACTAACGCCCTAGCGTAAGAACGAGCTAAATAGCGAGCGGCCT-
CAAGTGCTAACGTAAGAACACTCATCCAATAGAAATAGATCCAGTCC
AGGCTCCCGTGAGTCATAAATAGATAAATACATACAATCAAGGC-
TATCTCCCGAGAGAACGAGATAGCGAGAGAACTGTCTAGCGCCCT-
CACTCCCTAGCGTGAGAACTCCCGTGCTAACGACCGATCGGTAGGA
AGAGCATGAGAGAGTAAGAACTAACGACCGATCGGTAGGAAGAG-
CAAGATATATAGAGAGAGTAAGAACTATCTCCCGAGCTATCTCACTA-
GAGGAAGAGATAAAGAGAGAACTGAAGAAATATATAGAGAAAGTGA
GAACTATCGGACGGCCGCGCTCAAGTAAGAAAGTCATATATAGAG-
GCAGGCAGGCATGTCTTCAATCAAGGCGTTCTAACGCCCGTGA-
GAACATTCCCCCCCAAGTAAGAGATTGCATTCCCCCCCAAGAG

End of notes and DNA encoding notes # #

For more information, and to explore the possibility of getting your own
DNA copy of this book, please visit http://periodicplayground.com.

INDEX